Photograph Kobé, Wien I

Topics in Applied Continuum Mechanics

Symposium Vienna, March 1-2, 1974

Edited by
J. L. Zeman and F. Ziegler

Springer-Verlag Wien GmbH

With 1 Portrait and 67 Figures

Originally published by Springer-Verlag/Wien New York in 1974

Library of Congress Cataloging in Publication Data
Main entry under title:

Topics in applied continuum mechanics.

Papers presented at a symposium held at the University of Technology of Vienna in honor of H. Parkus.
Includes bibliographies.
1. Continuum mechanics--Congresses. 2. Parkus, Heinz, 1909- --Bibliography. I. Zeman, Josef L., 1938- ed. II. Ziegler, Franz, 1937- ed. III. Parkus, Heinz, 1909-
QA808.2.T63 620.1'05 74-12227

ISBN 978-3-211-81260-0 ISBN 978-3-7091-4188-5 (eBook)
DOI 10.1007/978-3-7091-4188-5

P R E F A C E

On January 31, 1974 Professor Dr.Heinz Parkus celebrated his 65th birthday. In his honour an international symposium on continuum mechanics, sponsored by the Austrian Secretary of Science and Research, was held on March 1 and 2, 1974 at the University of Technology of Vienna.

Here scientists presented their special fields of interest in a series of lectures, in reviews as well as in reports on latest developments in present day research, experimental as well as theoretical.

From these contributions those which seem to be of general interest are collected in this volume. They are arranged starting with experimental results, the basis of any phenomenological theory, followed by theoretical results, by results of application oriented theory and by results of applications which lead to experiments. Of course, none of the papers fits exactly into this scheme, some even cover the entire range themselves.

There is not space here to summarize the papers printed in this volume. We confine ourself to pointing out the remarkable experiments on plasticity by A.Phillips, the fundamental contribution of E.Kröner to self-stresses, the excellent review of electro-magneto-elasticity by J.B.Alblas, the theoretical treatise of distortion in micropolar elasticity by W.Nowacki, the theoretical considerations of J.A.König and W.Olszak on the influence of nonhomogeneity on plasticity, the two engineering oriented papers by W.R.Delameter-G.Herrmann and H.Bargmann covering some modern topics in elasticity, the reviews on plasticity and creep from a more engineering point of view by J.F. Besseling and J.Hult, and the outline of plastokinetics of metal forming by H.Lippmann.

A list of references, a subject index and a list of publications of Professor Heinz Parkus have been added. The list of references is complete and the underlined page numbers refer to pages where the complete references may be found. In the subject index only those pages are listed where definitions or general information may be found.

In concluding this preface we take the opportunity of expressing our sincerest thanks to the authors, who handed in the manuscripts in time and in a manner such that they could be forwarded to the publisher within short notice, and to Springer-Verlag Wien-New York for making it possible that this interesting collection of different views of the very same subject, applied continuum mechanics, can appear in such a short time.

J.L. Zeman F. Ziegler

C O N T E N T S

THE FOUNDATIONS OF THERMOPLASTICITY - EXPERIMENTS AND THEORY

A. PHILLIPS, New Haven

1. Introduction

Thermoplasticity deals with the interaction between temperature, stress and plastic deformation due to either a stress increment, or a temperature change, or to a combination of a stress increment and a change in temperature. As in the case of isothermal plasticity the concepts of the yield surface and of the plastic strain rate are of fundamental importance. The yield surface and the plastic strain rate are now functions of the temperature in addition to being functions of the other variables which appear in isothermal plasticity.

In this paper[1] we shall present some of the findings on the foundations of thermoplasticity which have been obtained by my research group at Yale University during the last few years. These findings include recently published material but also some not yet published results. In particular this paper will go much beyond the material which has been reviewed in the article of Professor J. BELL in the Handbook der Physik /1/.

---

[1]This research has been supported by the National Science Foundation of the United States government.

We shall consider a slowly changing temperature so that for all practical purposes we shall deal with a sequence of isothermal states. We shall also consider slowly varying stresses so that we have to deal with quasi-equilibrium cases both from the stresses and the temperature point of view.

The yield surface represents the locus of all points in the seven-dimensional stress-temperature space enclosing the region in which any motion of the representative stress-temperature point will produce elastic strain only. Any excursion of the generic point outside the yield surface will produce permanent strains. Figure 1 shows in the three-dimensional stress-temperature space typical intersections of yield surfaces by a τ-T coordinate plane where τ is a shearing stress and T is the temperature. We see that as the temperature increases the yield surface decreases in size in the stress direction. From Figure 1 it is obvious that plastic deformation will occur even when the stress does not change or when it decreases, provided that the temperature increases appropriately.

The concept of the yield surface is fundamentally connected with the definition of yielding and with the first appearance of plastic strain. If the yield surface is to be the region in stress-temperature space in which for any motion of the generic point only elastic strains appear it is necessary that the stress at yielding be defined as equivalent to the stress at the proportional limit, that is, to the stress at the first deviation from proportionality. This is the definition we shall adopt in this paper as we have consistently adopted in all of our work. It is of interest to note that this is not the definition which was always used by other authors even for the determination of yield surfaces[2].

In some recent accounts /2,3/ of the theory of plasticity the attempt was made to avoid having to make the assump-

---

[2]See, /1/ for a historical account of the definition of yielding.

tion that a yield surface exists. The authors of these papers prefer the theory to predict the existence of a yield surface instead of assuming its existence in the first instance. To this author, such an attempt seems to be of academic interest. We do know from experiments that yield surfaces exist and it is immaterial whether the yield surface is a derived or a primary quantity. Only the simplicity of a particular theory will decide whether the yield surface should be an assumed or a derived concept. Of course any proposed theory, simple or not, should agree with the available experiment results. In addition its validity will be the more secure the larger the number of predictions which are experimentally verified.

Following /4/ we introduce a function

$$F(\sigma_{k\ell}, \varepsilon''_{k\ell}, T, \kappa) = 0 \tag{1}$$

which represents the yield surface. Here $\sigma_{k\ell}$ and $\varepsilon''_{k\ell}$ represent the stress and plastic strain tensors respectively. The scalar κ depends on the history of loading. For a given value of κ and of $\varepsilon''_{k\ell}$, Eq. (1) represents the yield surface in the seven-dimensional stress-temperature space. We now introduce the well-known concepts of loading, neutral loading, and unloading by means of the following equations, respectively:

$$\begin{aligned}
&\frac{\partial F}{\partial \sigma_{k\ell}} \dot{\sigma}_{k\ell} + \frac{\partial F}{\partial T} \dot{T} > 0 \text{ , } F = 0 \; (\dot{\varepsilon}_{k\ell} \neq 0, \; \dot{\kappa} \neq 0) \\
&\frac{\partial F}{\partial \sigma_{k\ell}} \dot{\sigma}_{k\ell} + \frac{\partial F}{\partial T} \dot{T} = 0 \text{ , } F = 0 \; (\dot{\varepsilon}_{k\ell} = \dot{\kappa} = 0) \\
&\frac{\partial F}{\partial \sigma_{k\ell}} \dot{\sigma}_{k\ell} + \frac{\partial F}{\partial T} \dot{T} < 0 \text{ , } F = 0 \; (\dot{\varepsilon}_{k\ell} = \dot{\kappa} = 0), \text{ or} \\
&\qquad F < 0, \; (\dot{\varepsilon}_{k\ell} = \dot{\kappa} = 0)
\end{aligned} \tag{2}$$

The differentiations in (2) are carried out with κ and $\varepsilon''_{k\ell}$ held fixed, since F is a yield surface corresponding to given values of $\varepsilon''_{k\ell}$ and κ. When $\dot{\kappa} \neq 0$, $\dot{\kappa}$ is assumed to satisfy an equation of the form

$$\dot{\kappa} = \dot{\kappa}\,(\sigma_{k\ell}, \varepsilon''_{k\ell}, T, \dot{\sigma}_{k\ell}, \dot{\varepsilon}''_{k\ell}, \dot{T}) \tag{3}$$

which is homogeneous of degree one in $\dot{\sigma}_{k\ell}$, $\dot{\varepsilon}_{k\ell}$, and $\dot{T}$. It should also satisfy the requirement

$$\dot{\kappa} = 0 \quad \text{when} \quad \dot{\varepsilon}''_{k\ell} = 0 \qquad (4)$$

When $\dot{\varepsilon}_{k\ell} \neq 0$, $\dot{\varepsilon}_{k\ell}$ is assumed to satisfy an equation of the form

$$\dot{\varepsilon}''_{k\ell} = g_{k\ell}\,(\sigma_{mn}, \dot{\sigma}_{mn}, \varepsilon''_{mn}, T, \dot{T}) \qquad (5)$$

which is homogeneous of degree one in $\sigma_{k\ell}$ and $\dot{T}$. Equation (5) gives the plastic strain rate as a function of the variables indicated.

2. The Initial Yield Surface

In this section we shall consider the yield surface at a stage when no plastic deformation has been introduced into the specimen. We have, therefore, $\varepsilon''_{k\ell} = 0$ and $\kappa = 0$. The equation for the initial yield surface is

$$F(\sigma_{k\ell}, T) = 0 \qquad (6)$$

Figure 2 from /5/ shows for commercially pure aluminum annealed at 650°F the initial yield surface in the tension-torsion space. In this figure the yield surface is represented by means of its isothermals at the temperatures of 70°F, 151°F, 227°F, and 305°F. The same specimen was used to obtain all data points at all temperatures. All the data on the 70°F isothermal were determined before raising the temperature to 151°F and the same procedure was followed for the higher temperatures until all the data at the highest temperature were obtained. It will be observed that the curves shown are little different from those which would have been predicted by the Mises theory. We have also found in additional experiments that even the order in which the temperature was changed is immaterial. In the σ, τ, T space the yield surface is a truncated elliptical cone between the temperatures of 70°F and 305°F. The left part of Fig. 1 shows the intersection between the elliptical cone and the τ-T plane. We see that this intersection consists of two straight lines.

Figure 3 from /5/ shows our results for yield surfaces in the σ_y, σ_z space at different temperatures for pure aluminum. The similarity with the Mises surface is very clear. Figure 4 from /5/ shows the three-dimensional yield surface for aluminum in the σ_y, σ_z, τ_{yz} space at room temperature. It has been constructed partly from Figs. 2 and 3 and partly from additional data.

In the above discussion we considered commercially pure aluminum. Now we turn our attention to two other materials, copper and brass. Figure 5 shows the initial yield surface at room temperature in the σ, τ space for oxygen free high conductivity copper which was annealed for one hour at 750°F in a vacuum furnace. Figure 6 shows the initial yield surface at room temperature in the σ_z, σ_y space[3] for the same material as above. From Figs. 5 and 6 we observe that copper follows more closely the Tresca surface than the Mises surface.

Figure 7 shows the initial yield surfaces for brass annealed at 1080°F for one hour. The yield surface is shown in the σ, τ plane and it has been obtained at 75°F and at 200°F. It agrees with the Tresca condition.

A novel way to determine yield surfaces is by means of increasing the temperature while the stress remains constant, increases, or even decreases. Figure 1 shows some of the possible paths which can be used. In /6/ this method has been used and we observed that the intersections between the yield surface and the non-isothermal paths which we obtained coincide with the prediction from the isothermal tests.

From the experimental results discussed above we can conclude that when $\kappa = 0$ and $\varepsilon''_{k\ell} = 0$ and the material is annealed the function F in Eq. (1) will have the form

$$F(J'_2, J'_3, T) = 0 \tag{7}$$

where the relation between the stress and the temperature T is a linear one. In Eq. (7) J'_2 and J'_3 are the second and

---

[3] In Figure 6 σ_y is denoted as σ_θ .

third invariants of the deviators of stress. The influence of J_3' on the function F is minimal for aluminum but substantial for copper and brass.

3. The Subsequent Yield Surface

Prestressing changes the yield surface. We obtained experimentally a large number of subsequent yield surfaces due to prestressing. Most of these surfaces were for aluminum specimens and a smaller number were for copper and brass. Let us give some examples.

Figure 8 from /7/ shows a sequence of two prestressings for aluminum. In this case the initial yield surface was obtained first, then the specimen was prestressed to the point A and then the first subsequent yield surface was obtained. Subsequently, the loading path starting from inside the first subsequent yield surface proceeded to prestress the same initial specimen to the point B and then the second subsequent yield surface was obtained. All prestressings were at room temperature. The isothermals at four temperatures were obtained for the initial yield surface and for the first subsequent yield surface. For the second subsequent yield surface the isothermals at only three temperatures were obtained.

From the above described test we may draw a number of important conclusions which have been verified by a large number of additional experiments. We first observe that the yield surface does not pass through the prestressing point. Whether the yield surface will pass very near the prestress point or not depends on the time elapsed while the specimen is held at the prestressing point. If the specimen after being prestressed is allowed to remain at the prestressing point for considerable time the yield surface will pass very near the prestressing point. If, on the other hand, after prestressing the stress is immediately reduced to a value within the corresponding yield surface, the yield surface will be at considerable distance from the prestressing point.

An elementary intuitive explanation of this phenomenon is

given in /8/. It is based on the idea that in the process of prestressing with a given stress rate the permanent strain corresponding to a given stress does not have sufficient time to develop. As explained in /8/ let for simple tension, Fig. 9, line OA be the equilibrium stress-strain line, i.e. the lowest stress-strain curve possible at a given temperature. This is a curve in which by definition after every stress increment $d\sigma$ was applied sufficient time lapsed for all the permanent strain to have appeared before a new stress increment was applied. Let also OB be a stress-strain line obtained with some finite stress rate. If, while obtaining the curve OB, the stress remained constant at the point E, then the strain would have continued to increase until the equilibrium curve OA would have been reached at A with a final plastic strain ε_A . If, however, upon reaching E we decrease the stress to a point C below the equilibrium line crossing the line at D, we freeze the amount of plastic strain accumulated to the amount ε_D, which has been reached while crossing the equilibrium line at D. Reloading again we obtain line CDF where CD is the elastic reloading line coincident with the elastic unloading line while DF does not coincide with ED. Thus the point D of the equilibrium line corresponds to the prestress point E and to the plastic strain ε_D . If unloading had started at E, but after some additional plastic strain had appeared, for example at G, and the equilibrium had been crossed at H, the plastic strain would have been frozen at H, with the plastic strain value ε_H, which is different from ε_D . This is due to the time-development of plastic strain. In combined stress the axis σ in Fig. 9 is replaced by the stress space and the stress at the points D or H corresponds to points on the yield surface, while the stress at the points E or G corresponds to the prestress point.

The above interpretation is reinforced by the experimental result shown in Fig. 10 from /5/ where the subsequent yield surface was determined once and then the same prestressing point A was reached again and a new subsequent yield surface was determined. We see that the second subsequent yield surface, due to prestressing to the same point A as the first subsequent yield surface, lies in a position closer to the prestressing point than the first subsequent yield surface, since additional plastic strain was developed. The amount of plastic strain developed while the loading point is outside the yield surface determines the closeness of the yield surface to the prestressing point. It is seen from Fig. 10 that the change in the yield surface during the period in which the loading point remains in the prestressing position does not affect the size of the yield surface in any direction lateral to the prestressing direction.

Let us return now to the experiment shown in Fig. 8. A second observation of importance is that the subsequent yield surfaces do not include the origin. For torsional loading only, this phenomenon has also been observed by Ivey /9/.

A third observation is that prestressing moves the yield surface in such a manner that cross effect does not occur. In a more general form this lack of cross effect can be explained on the basis of Fig. 11 from /5/. Suppose that at some constant temperature T and at some level of prestressing the yield surface in the σ_y, σ_z, τ stress space is given by S_i and that the stress path which is responsible for the surface S_i terminates at the loading point P_i. We retrace the stress path leading to P_i backwards until we reach some arbitrary position R_i inside S_i. Since any motion of the loading point inside or on S_i will not alter S_i let us select a new arbitrary position Q_i inside or on S_i not necessarily on the stress path leading to P_i. Suppose now that additional prestressing is generated by an arbitrary rectilinear motion of the loading point from Q_i inside S_i to a position P_{i+1} outside S_i. The position P_{i+1} may of course be the same as P_i. Then the new yield surface S_{i+1} corresponding to the stress path terminating at P_{i+1}, is generated from the surface S_i by a superposition of a rigid body translation in the direction of prestressing Q_iP_{i+1} and of a deformation in the same direction Q_iP_{i+1}. The effect of the deformation is that the width of the yield surface in the direction of prestressing will decrease. In particular, the motion of the forward part of the yield surface will usually be less than the motion of its rear part.

The amount of the rigid body translation is determined by the motion of the curve ABCD to its new position A'B'C'D'. The motion of this curve generates a cylinder with its axis in the direction of prestressing. This cylinder is tangential to both the original and to the new yield surface at the curves ABCD and A'B'C'D', respectively.

A preliminary analytical formulation of this law appeared in /10/.

In Fig. 12 from /5/ we see the σ, τ plane for an experiment in which the three dimensional yield surfaces for an aluminum specimen were obtained after three successive prestressings in torsion, tension, and torsion, respectively. In Fig. 13 from /5/ the projections of the yield surfaces on the σ_y, σ_z plane are given. From both figures we see that our hardening law is valid. In Fig. 14 from /5/ is shown the three dimensional form of the first subsequent yield surface of the specimen shown in Figs. 12 and 13. It is remarkable that the law of hardening we proposed is valid for all yield surfaces at all tested temperatures and for prestressing at any temperature.

Figures 15, 16 and 17 from /11/ present the case of an aluminum specimen in which the prestressing was of a cyclic nature. The specimen was prestressed first to A, Fig. 15, then to B, Fig. 16, and finally to C, Fig. 17. The hardening law is valid and as shown from Fig. 18 the width of the yield surface in the direction of prestressing first decreases, then increases, and then decreases again.

Figure 19 shows the motion of the yield surface for a copper specimen for five consecutive prestressings. Again our law of hardening is shown to be correct. This particular figure shows also the increments of plastic strain at the points where the subsequent prestressing path crosses the previous yield surface. It is seen that normality between the yield surface and the strain rate vector is valid. This observation has been made in all our tests.

4. Some Theoretical Consequences

The results given in the previous sections lead to some interesting theoretical consequences. Let us assume that the free energy function A and the entropy S are given by

$$A = A(\varepsilon'_{k\ell}, \varepsilon''_{k\ell}, \kappa, T) \tag{8}$$

$$S = S(\varepsilon'_{k\ell}, \varepsilon''_{k\ell}, \kappa, T) \tag{9}$$

where $\varepsilon'_{k\ell}$ is the elastic strain. If $\varepsilon_{k\ell}$ is the total strain then we also assume that

$$\varepsilon_{k\ell} = \varepsilon'_{k\ell} + \varepsilon''_{k\ell} \tag{10}$$

since we deal with small strains only. As shown in /4/ and /12/ the laws of thermodynamics give

$$S = -\frac{\partial A}{\partial T} \tag{11}$$

$$\sigma_{k\ell} = \rho_o \frac{\partial A}{\partial \varepsilon'_{k\ell}} \tag{12}$$

provided that the heat flux vector Q_k is given by

$$Q_k = Q_k(\varepsilon'_{k\ell}, \varepsilon''_{k\ell}, \kappa, T, T_{,m}) \tag{13}$$

and that the temperature distribution is homogeneous. In addition we obtain

$$\left(\sigma_{k\ell} - \rho_o \frac{\partial A}{\partial \varepsilon''_{k\ell}}\right) \cdot \dot{\varepsilon}''_{k\ell} \geqq 0 \tag{14}$$

where $\sigma_{k\ell}$ is any arbitrary point of the yield surface. The quantity ρ_o $(\partial A/\partial \varepsilon''_{k\ell})$ has the dimensions of stress and will be denoted by

$$\rho_o \frac{\partial A}{\partial \varepsilon''_{k\ell}} = s^o_{k\ell} \tag{15}$$

In /13/ it has been called the "thermodynamic reference stress."

In Eq. (12) the left hand term can be determined from the elastic strain and the temperature. Hence $\sigma_{k\ell}$ is not a function of the plastic strain or of the history of plastic strain. Thus $\partial A/\partial \varepsilon'_{k\ell}$ must also be independent of κ and $\varepsilon''_{k\ell}$. We can therefore write

$$A = A'(\varepsilon'_{k\ell}, T) + A''(\varepsilon''_{k\ell}, \kappa, T) \tag{16}$$

Then, equation (11) gives

$$S = S'(\varepsilon'_{k\ell},\ T) + S''(\varepsilon''_{k\ell},\ \kappa,\ T) \tag{17}$$

where

$$S'(\varepsilon'_{k\ell},\ T) = -\ \frac{\partial A'(\varepsilon'_{k\ell},\ T)}{\partial T}\ , \tag{18}$$

$$S''(\varepsilon''_{k\ell},\ \kappa,\ T) = -\ \frac{\partial A(\varepsilon''_{k\ell},\ \kappa,\ T)}{\partial T} \tag{19}$$

During unloading, when $d\varepsilon''_{k\ell} = 0$, $d\kappa = 0$, dS should be independent of the particular value of $\varepsilon''_{k\ell}$ and κ . Hence

$$dS = \frac{\partial S'}{\partial \varepsilon'_{k\ell}}\, d\varepsilon'_{k\ell} + \frac{\partial S'}{\partial T}\, dT + \frac{\partial S''}{\partial T}\, dT \tag{20}$$

It follows that $\partial S''/\partial T = c$ = constant independent of $\varepsilon''_{k\ell}$, κ, and

$$S'' = cT + c_1(\varepsilon''_{k\ell},\ \kappa)\ . \tag{21}$$

Now in

$$S = S'(\varepsilon'_{k\ell},\ T) + cT + c_1(\varepsilon''_{k\ell},\ \kappa) \tag{22}$$

we can incorporate cT into $S'(\varepsilon'_{k\ell},\ T)$ and thus write

$$S'' = c_1(\varepsilon''_{k\ell},\ \kappa) = S''(\varepsilon''_{k\ell},\ \kappa) \tag{23}$$

and from (19) we obtain

$$\frac{\partial A''}{\partial T} = -\ S''(\varepsilon''_{k\ell},\ \kappa) \tag{24}$$

It follows

$$A'' = -\ T\ S''(\varepsilon''_{k\ell},\ \kappa) + A'''(\varepsilon''_{k\ell},\ \kappa) \tag{25}$$

from which we obtain

$$A = A'(\varepsilon'_{k\ell},\ T) - T\ S''(\varepsilon''_{k\ell},\ \kappa) + A'''(\varepsilon''_{k\ell},\ \kappa) \tag{26}$$

Equations (14), (17) and (26) are fundamental for our considerations. Equation (26) has been developed in /14/ by a different method.

Now Eq. (14) becomes

$$\left(\sigma_{k\ell} - \rho_o\,\frac{\partial A''}{\partial \varepsilon''_{k\ell}}\right)\cdot\ \dot{\varepsilon}''_{k\ell} \geqq 0 \tag{27}$$

and

$$s^{o}_{k\ell} = \rho_o\,\frac{\partial A''}{\partial \varepsilon''_{k\ell}} = \rho_o\,\frac{\partial A'''}{\partial \varepsilon''_{k\ell}} - \rho_o\ T\,\frac{\partial S''}{\partial \varepsilon''_{k\ell}} \tag{28}$$

Here both $\rho_o \frac{\partial A'''}{\partial \varepsilon''_{k\ell}}$ and $\rho_o T \frac{\partial S''}{\partial \varepsilon''_{k\ell}}$ represent stresses.

At any particular values of $\varepsilon''_{k\ell}$ and κ the yield surface is determined, and at an arbitrary but particular value of the temperature, inequality (27) together with the normality rule mean that the stress point $s^o_{k\ell}$ corresponding to that particular temperature must lie within the corresponding isothermal yield curve. The higher the temperature becomes the smaller will be the area enclosed by the isothermal yield curve and the stress $s^o_{k\ell}$ will tend to be restricted towards the center of the cluster of isothermals.

In particular we shall have inequality

$$F(s^o_{k\ell}, \varepsilon''_{k\ell}, T, \kappa) < 0 \tag{29}$$

which shows that we should be careful when selecting the functions F, κ, $\dot{\varepsilon}''_{k\ell}$, A'', A''', and S'' so that inequality (29) will not be violated for any possible path of loading. An example where a deliberate selection of these functions leads to a violation of inequality (29) is given in /13/.

In Eq. (28) we shall write

$$\rho_o \frac{\partial A'''}{\partial \varepsilon''_{k\ell}} = s^{o'}_{k\ell} \tag{30}$$

and

$$\rho_o T \frac{\partial S''}{\partial \varepsilon''_{k\ell}} = s^{o''}_{k\ell} \tag{31}$$

so that we obtain

$$s^o_{k\ell} = s^{o'}_{k\ell} - s^{o''}_{k\ell} \tag{32}$$

In the six dimensional stress space Eq. (32) is a vectorial equation. We observe that $s^{o'}_{k\ell}$ is independent of T and therefore it is probably located at the center of the cluster of isothermals, Fig. 20. It could be assumed that as T increases $s^o_{k\ell}$ will tend to reach $s^{o'}_{k\ell}$. This is, however, not correct. Indeed, the vector $s^{o''}_{k\ell}$ is the product of the scalar $\rho_o T$ and the vector $\partial S''/\partial \varepsilon''_{k\ell}$. Now this second vector is a function of $\varepsilon''_{k\ell}$ and κ only, hence it is independent of the temperature and is determined by the yield surface. An increase in temperature tends to move $s^o_{k\ell}$ away from $s^{o'}_{k\ell}$.

Since, however $s^{o}_{k\ell}$ at each temperature must be inside the corresponding yield curve and the yield curves tend to converge towards the center $s^{o'}_{k\ell}$, it follows 1) that $s^{o}_{k\ell}$ must always lie within the innermost yield curves irrespective of the value of the temperature and 2) that the cluster of yield curves will never be reduced to a point except, of course, when $\varepsilon''_{k\ell} = 0$, $\kappa = 0$.

We now remark that $s^{o''}_{k\ell}$ is a vector of constant direction. Hence as T increases to the value it has during annealing the point $s^{o}_{k\ell}$ must always move in a rectilinear motion away from $s^{o'}_{k\ell}$ until it reaches its final position. Therefore, the yield curves will reduce in the limit to a straight line from $s^{o'}_{k\ell}$ to the final position of $s^{o}_{k\ell}$. By inspecting the yield curves in Fig. 8 it becomes clear that this is what happens as the temperature increases for any values of $\varepsilon''_{k\ell}$ and κ. We also observe that since the law of hardening obtained previously is independent of the temperature, it follows that as the yield surface moves in space because of prestressing the point $s^{o'}_{k\ell}$ will move parallel to the prestressing direction. This motion of $s^{o'}_{k\ell}$ is due to the change of $\partial A''/\partial\varepsilon''_{k\ell}$ with $\varepsilon''_{k\ell}$ and κ. In addition, the rotation of the yield surface is expressed by the rotation of the vector $s^{o''}_{k\ell}/\rho_o T$ which is due to the change of $\partial S''/\partial\varepsilon''_{k\ell}$ with $\varepsilon''_{k\ell}$ and κ.

5. References

1. BELL, J.: The experimental foundations of solid mechanics, in Handbook der Physik vol. VI A/1, pp. 679-684, Berlin: Springer 1973.
2. VALANIS, K.C.: A theory of viscoplasticity without a yield surface. Parts I and II. Archives of Mechanics, 23, 517-552, (1971).
3. FOX, N.: On the derivation of the constitutive equations of ideal plasticity, Acta Mechanica 7, 248-251, (1969).
4. GREEN, A.E. and NAGHDI, P.M.: A general theory of an elastic-plastic continuum. Archive for Rational Mechanics and Analysis, 18, 251-281, (1965).

5. PHILLIPS, A. and KASPER, R.: On the foundations of thermoplasticity - an experimental investigation. Journal of Applied Mechanics, Transactions of the ASME, ser. E, 40, 891-896, (1973).
6. PHILLIPS, A., LIU, C.S., and JUSTUSSON, J.W.: An experimental investigation of yield surfaces at elevated temperatures. Acta Mechanica, 14, 119-146, (1972).
7. PHILLIPS, A. and TANG, J.L.: The effect of loading path on the yield surface at elevated temperatures. International Journal of Solids and Structures, 8, 463-474, (1972).
8. PHILLIPS, A.: Experimental plasticity. Some thoughts on its present status and possible future trends. General lecture at International Symposium on Foundations of Plasticity. To appear in the Proceedings, Vol. II. (A Sawczuk, editor), Noordhoff (1973), (in press).
9. IVEY, H.J.: Plastic stress-strain relations and yield surfaces for aluminum alloys. Journal of Mechanical Engineering Science, 3, 15-31, (1961).
10. PHILLIPS, A.: New theoretical and experimental results in plasticity and creep. Proceedings 2nd International Conference on Structural Mechanics in Reactor Technology, 5, L 3/6, (1973).
11. PHILLIPS, A., TANG, J. L. and RICCIUTI, M.: Some new observations on yield surfaces. Acta Mechanica. (in press).
12. GREEN, A. E. and NAGHDI, P.M.: A thermodynamic development of elastic-plastic continua, IUTAM Symposium on Irreversible Aspects of Continuum Mechanics. Wien, Springer (1968), pp. 117-131.
13. PHILLIPS, A. and EISENBERG, P.M.: Observations on certain inequality conditions in plasticity. International Journal of Non-linear Mechanics, 1, 247-256, (1966).
14. HADDOW, J. B.: A note on the thermodynamics of elastic-plastic deformation. Journal of Applied Mechanics 38, 541-543 (1971).

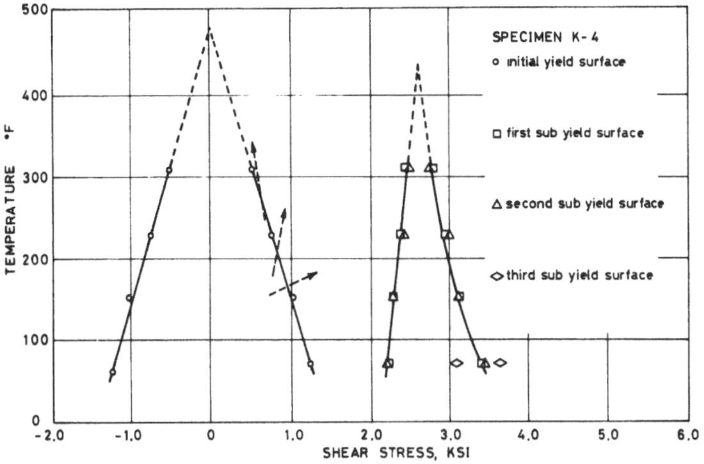

Fig. 1. Cross-Sections of Yield Surfaces.

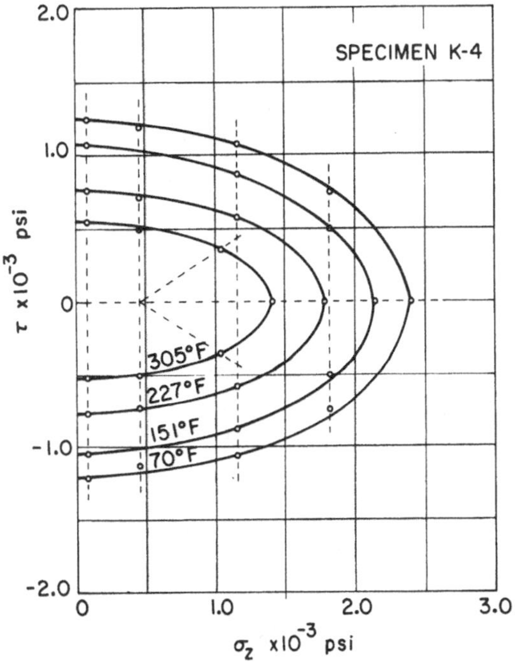

Fig. 2. Initial Yield Curves. Aluminum.

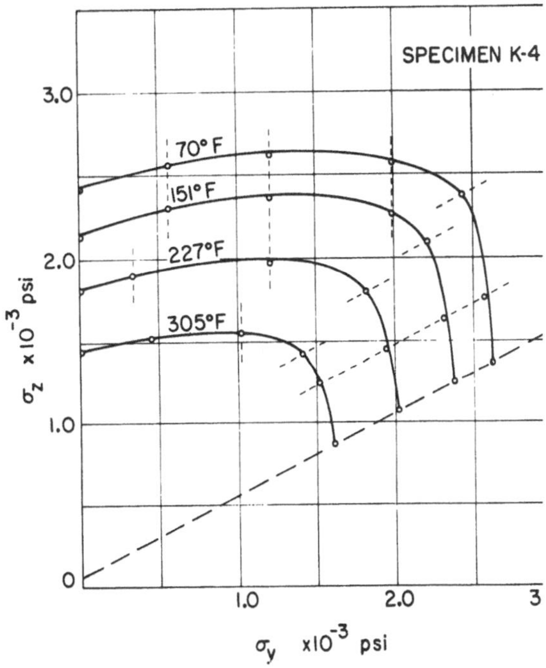

Fig. 3. Initial Yield Curves. Aluminum.

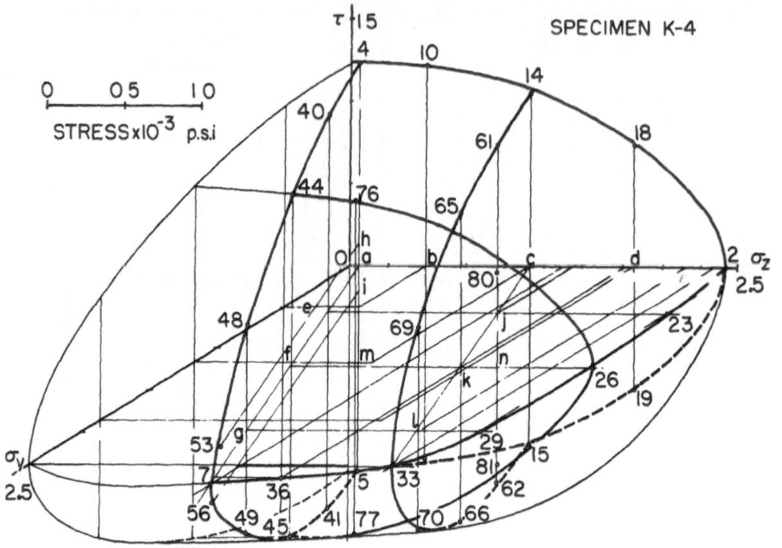

Fig. 4. Initial Yield Surface. Aluminum.

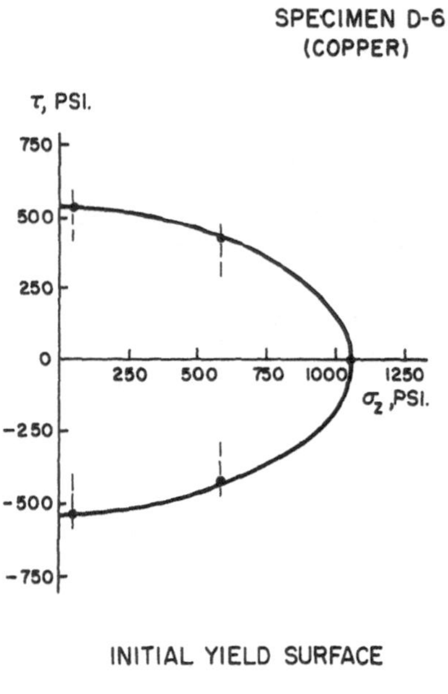

Fig. 5. Initial Yield Curve. Copper.

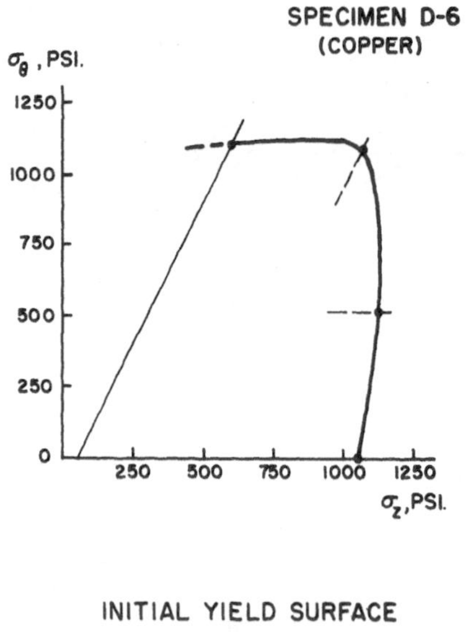

Fig. 6. Initial Yield Curve. Copper.

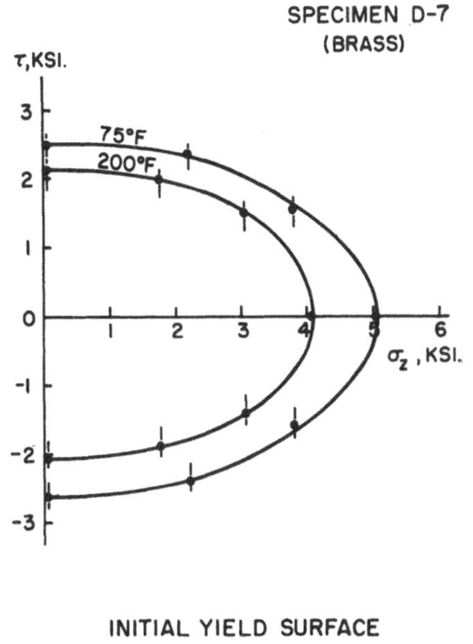

Fig. 7. Initial Yield Curve. Brass.

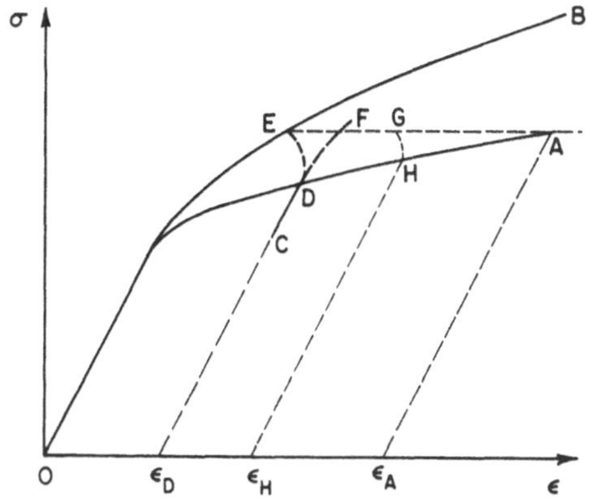

Fig. 9. The Equilibrium Stress-Strain Line.

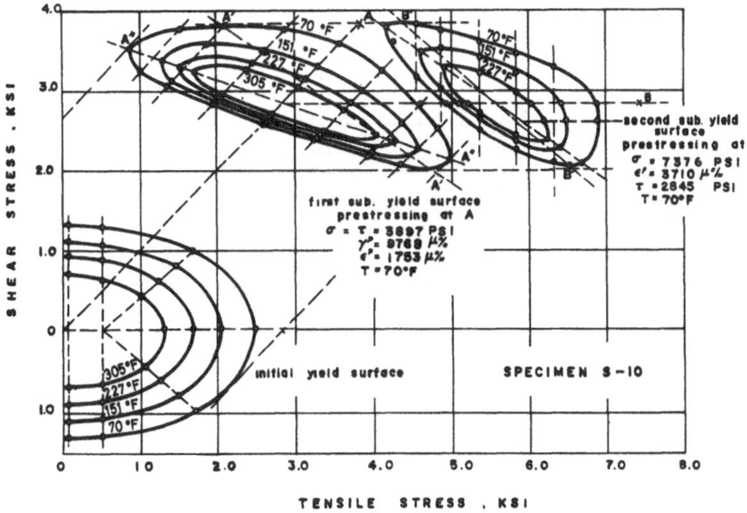

Fig. 8. Subsequent Yield Curves. Aluminum.

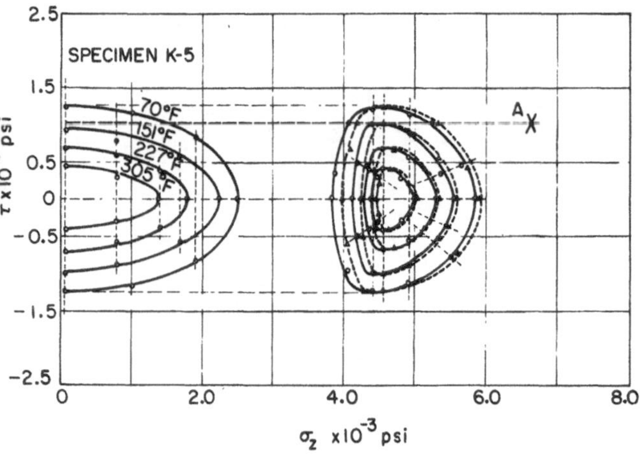

Fig. 10. Repeated Loading. Aluminum.

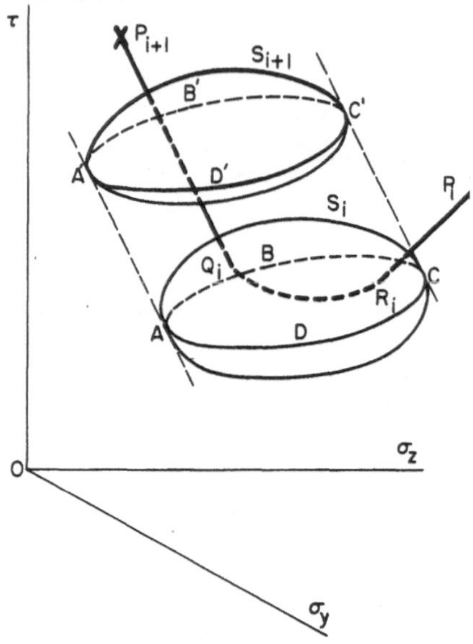

Fig. 11. The Hardening Law.

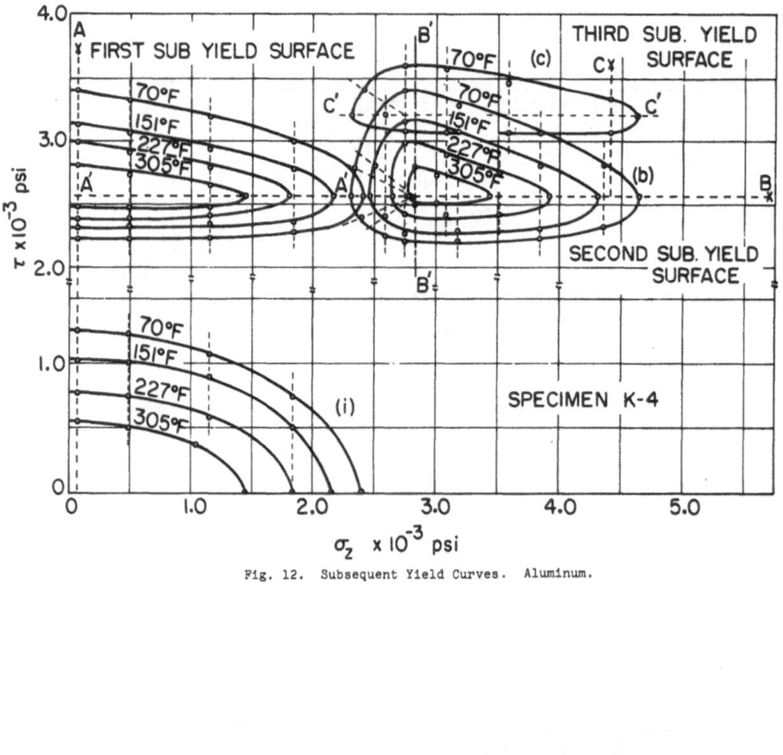

Fig. 12. Subsequent Yield Curves. Aluminum.

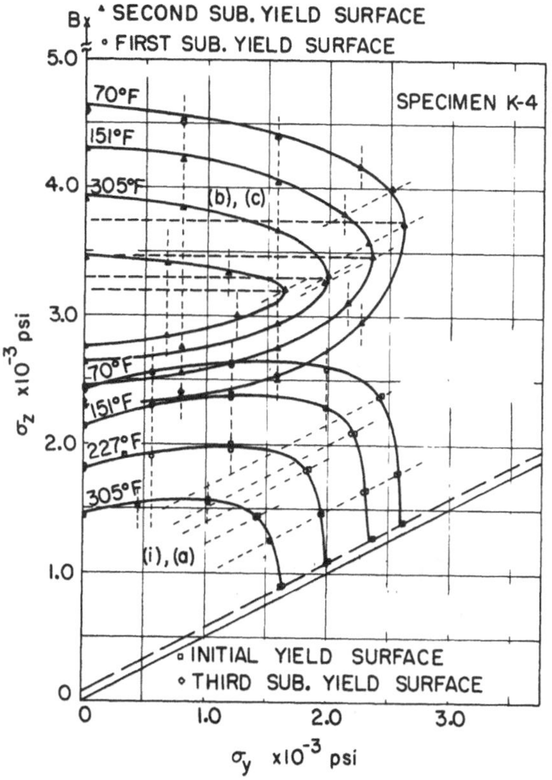

Fig. 13. Subsequent Yield Curves. Aluminum.

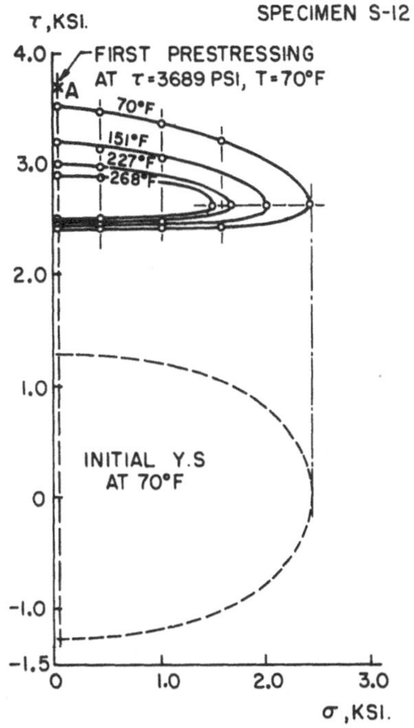

Fig. 15. Cyclic Loading. Part I. Aluminum.

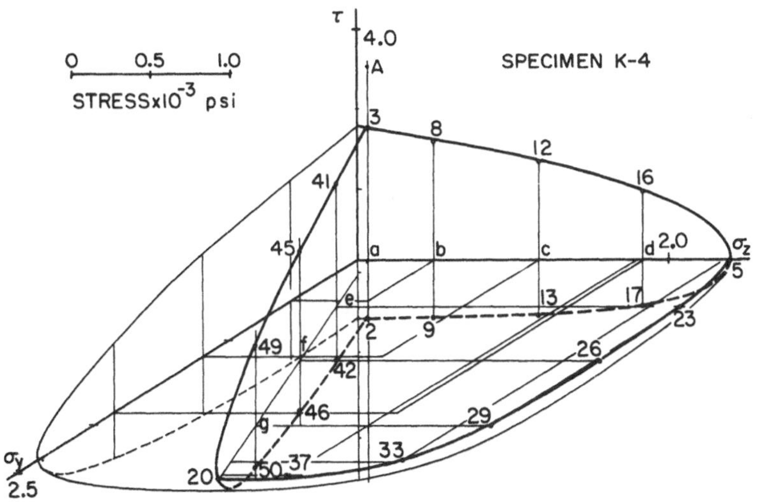

Fig. 14. Subsequent Yield Surface. Aluminum.

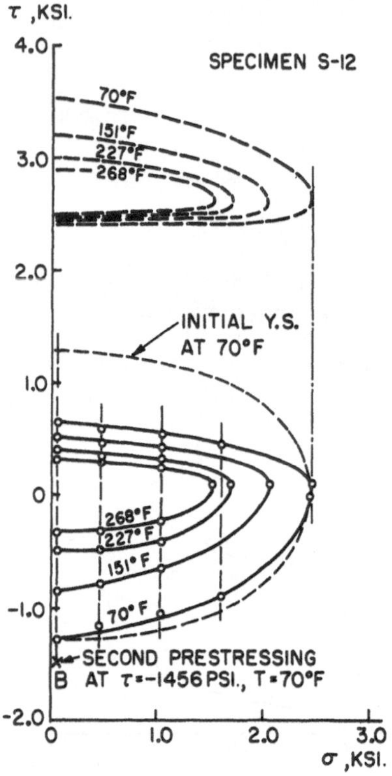

Fig. 16. Cyclic Loading. Part II. Aluminum.

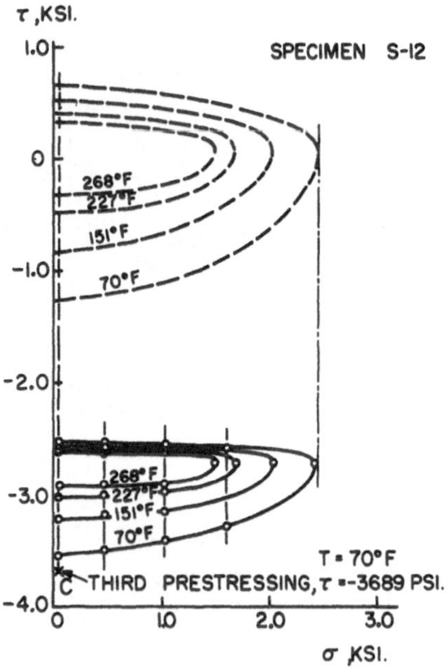

Fig. 17. Cyclic Loading. Part III. Aluminum.

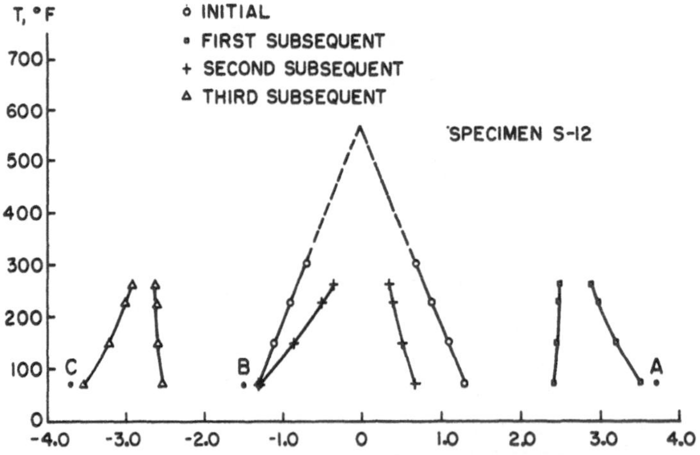

Fig. 18. Cyclic Loading. Cross-section.

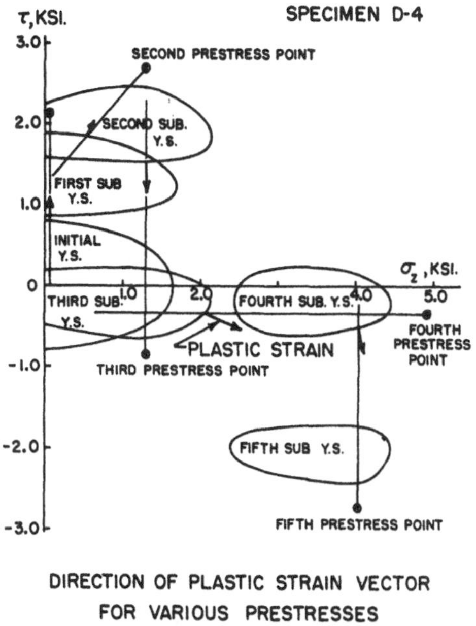

Fig. 19. Motion of Yield Surface. Copper.

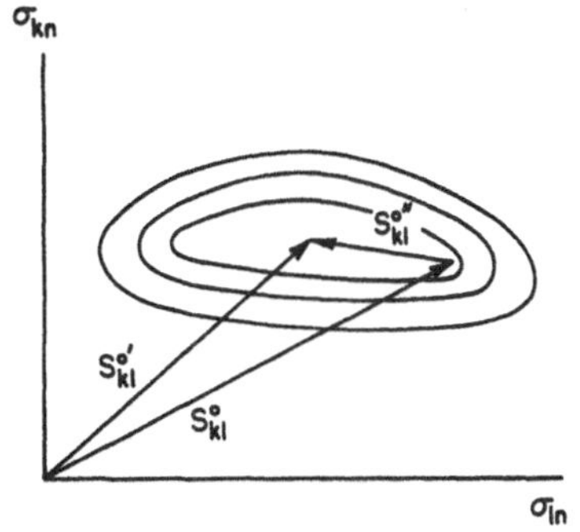

Fig. 20. The Three Vectors $s^{o}_{k\ell}$.

ON THE PHYSICS AND MATHEMATICS OF SELF-STRESSES

E. KRÖNER, Stuttgart

1. Introduction

By self- or residual stresses we shall understand those stresses which in a solid material remain after any kind of non-elastic treatment such as plastic deformation, heating and cooling, recrystallization, phase transformation etc. These stresses can exist without the action of external forces.

The equations of non-linear elasticity theory admit further kinds of stresses which occur in the absence of external forces. The stresses in an inverted shell are a typical example. These stresses will not be considered here.

Stresses which stem from sources other than mechanical can also exist without the action of external forces. Particularly often treated are thermal and magnetostrictive stresses. They, too, will be excluded from the notion of self-stresses employed here although the methods to be discussed will suit well to many problems involving such stresses.

In section 2 a few sources of self-stresses will be considered. These are the most important and elementary sources of self-stresses in pure materials, i.e. in those materials which consist of particles of one kind only. We shall think of atoms, but they could be molecules just as well. The materials of this kind correspond to the materially uniform bodies of the mathematical theory of elasticity.

Generally speaking, the sources of self-stresses are defects in an otherwise regular structure of which the crystal is the best-understood example. The lattice defects have been the object of scientific research for decades and have been found to be responsible for a huge variety of properties of such bodies to which among others all metals belong. Therefore, the account given here cannot be exhaustive by far. It rather concentrates on very few characteristic examples.

In section 3 we outline how the physical sources of self-stresses can be described in a macroscopical mathematical language. The basic equations are formulated which govern the mathematical problem of the self-stresses. A powerful tool for the solution of such problems is the method of the modified Green's functions which is explained in section 4 and is the central part of this lecture. A valuable feature of the method consists in the fact that it can be given in a most compact form and allows to write down general solutions in an impressively clear form. The method suits equally to problems with external stresses and takes into account the boundary conditions of the problem from the very beginning.

Section 4, finally, contains some concluding remarks.

2. The Physical Origin of the Self-Stresses

According to a fundamental law of nature a number of particles-atoms, molecules etc.- which are in thermal equilibrium and form a solid body assume that global configuration for which the free energy takes the minimum value. As is well-known the free energy contains a contribution of entropy which increasingly gains importance as the temperature rises. As a consequence, the particles do not assume the configuration of a perfect space lattice when they are in thermal equilibrium. This is true even if all particles are of the same sort. On the other hand, the internal energy strongly favours the formation of such crystal lattices. Therefore, the deviations from the ideal lattice are quite small in many solids. We shall be concerned mainly with such materials.

The most important deviations from the perfect crystal structure, as far as the thermal equilibrium is concerned, are the so-called point-defects which we also call point-eigen-defects because we have confined ourselves to pure materials, i.e. we have excluded the formation of defects by external sources such as atoms of a different species. Of particular interest are the vacancy and the interstitial which are shown in Fig. 1a,b.

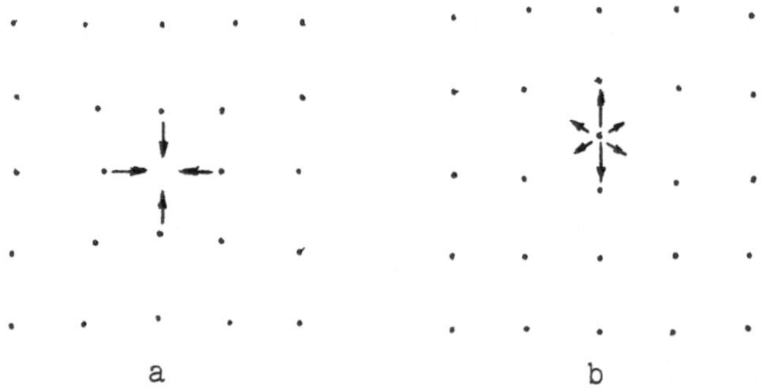

Figur 1

a Vacancy

b Interstitial

c Dislocation

If the energy of formation of these defects is known then it is possible to calculate their concentration in equilibrium as a function of temperature. A typical value near the melting point is 0.01%. The number decreases towards lower temperature.

It is another question whether the equilibrium is really achieved. This can only happen when the defects can move in the crystal. Since the motion of point-defects needs thermal activation the motion dies out towards low temperatures. To give an example: vacancies become immobile in copper at about

60 C whereas interstitials which occur in much lower concentration can move below room temperature. We see that the equilibrium concentration of point-defects is frozen in when the specimen is cooled to sufficiently low temperatures.

In a crystal which has no internal surfaces and no dislocations most of the vacancies and interstitials will migrate from the external surface. The equilibrium concentration can be changed at a fixed temperature by applying a hydrostatic pressure, for instance. This effect is reversibel, hence elastic, if the experiment is performed very slowly. The crystal diminishes its volume not only by shortening the atomic distances but in addition by giving away some of the space consuming vacancies and drawing in additional space saving interstitials. The latter effects can be called an elastic or, if one prefers, anelastic deformation. This is an interesting example of a global deformation which cannot be described as a superposition of local deformations. The defect itself which comes and goes is neither a strain nor a deformation. But the totality of defects in a larger region of the crystal has the status of a strain.

It appears plausible and agrees with theoretical and experimental results of solid state physics that the particles around a vacancy relax somewhat towards the defect whereas the particles around an interstitial are repelled. This deformation effect of the defect can be described approximately by the displacement field of, in general, a triaxial double force without moment as shown in Fig.1. Since the decay of the displacement field goes as $1/r^2$ with distance r it encompasses a large number of particles. It is therefore sensible to describe, at least in some distance from the defect, the atomic lattice as an elastic continuum to which stress and strain tensor fields can be assigned. Reflections of this kind establish the great importance of the elasticity theory also for problems of solid state physics where most investigations are done on a microscopic or submicroscopic scale.

The stresses and strains around a vacancy or interstitial decay as $1/r^3$. They exist without the action of external forces and are self-stresses in our present meaning. Apparently, it is sensible to understand these defects as elementary sources of the self-stresses.

We now introduce a further lattice defect, the often-quoted dislocation. In contrast to the point-defects which are also classified as 0-dimensional because they do not extend in any direction the dislocation has a line-form, i.e. it is a 1-dimensional quantity. This is shown in Fig. 1c. We shall not discuss here the interesting distinction between the edge and screw dislocation because we only discuss properties which are common to these two types of elementary dislocations.

It can be expected from the configuration of the dislocation that it is mobile through the lattice thereby causing a permanent relative shearing of the two parts of the crystal which are separated by the glide plane, the plane of the moving dislocation. The phenomenologically observed plastic deformation of crystalline bodies is nothing else than the superposition of billions of glide acts of dislocations.

It is the line-shape of the dislocations which is responsible for the fact that dislocation distributions do not form themselves in thermal equilibrium. They rather are produced by non-equilibrium processes such as crystallization and plastic deformation. A typical value for the number of dislocations piercing through an area element of $1cm^2$ in a weakly deformed metal-crystal is 10^8. During the crystal growth and later treatments the dislocations become interconnected in a 3-dimensional network. The nodes of this network are immobile for energetical reasons and therefore prevent the dislocations from escaping the crystal. Thus the crystal cannot pass to the configuration of equilibrium, i.e. the configuration of minimal free energy. Instead, it remains in a state of inhibited equilibrium which, however, varies if the dislocation distribution and number changes during plastic deformation. From a mechanical standpoint every crystal with dislocations is in a metastable equilibrium state. The complete specification of this state requires the knowledge of the line course and the type of each dislocation.

Therefore, great caution should be exercised in applying the thermodynamics of reversible or irreversible processes to problems of dislocations, in particular to plastic deformation.

Like the point defects dislocations too surround themselves with fields of stress and strain which in the case of a straight dislocation in an infinitely extended crystal decay as $1/r$ with distance r from the dislocation. Apparently we have again a self-stress situation. The sources are now the dislocations. Just as before it is justified to classify distributions of dislocations as global strains. This point is emphasized in the so-called continuum or field theory of dislocations.

3. Formulation of the Mathematical Problem of Self-Stresses

In the last section we have sketched the picture of a solid which has a crystalline structure and is filled with self-stresses. The elementary sources of these stresses have an extension of the order of one atomic distance in two or three directions. Due to the large range of the stresses it is sensible to make the transition to the continuum. Hereby the single sources assume the character of delta functions. In this manner a typical field theory is obtained with all the known difficulties and beauties.

Perhaps the most important problem consists in the task of calculating the sequence of the macroscopically registered stress and strain states, given the time sequence of the external loading. Of course, this task, when compared with the common elasticity theory, is non-trivial only if the deformation is plastic. In that case additional sources of self-stresses are created and displaced together with the already existing ones. This is no longer a purely mechanical problem since many processes with lattice defects can occur only with the help of thermal activation and therefore require a wider framework. In spite of many efforts in this direction a real break-through has yet to be achieved.

The problem would essentially be solved if one succeeded

to determine the temporal and spatial development of the sources of the self-stresses. An additional problem in this connection is the calculation of the interaction of the lattice defects among themselves and with the surface of the body. This problem requires the calculation of self-stresses the sources of which are given functions of position and perhaps of time. If the sources are near the surface we are faced with a boundary value problem.

A powerful tool of field theory, in particular in the case of general reflections, is the method of Green's functions. The direct application of this method is restricted to linear problems. However, non-linear problems can often be solved by an iterative treatment of linear equations. Hence the Green's method actually has an extraordinarily wide range of application.

REISSNER /1/ has pointed out in 1931 that self-stresses are always associated with incompatible elastic strains ε^{el}_{ij}. If we write de St.Venant's compatibility equations in the short hand notation[1]

$$\text{Ink}\ \underline{\varepsilon}^{el}(\mathfrak{r}) = 0 \qquad (\varepsilon_{ikm}\varepsilon_{jln}\partial_l\partial_k\varepsilon^{el}_{mn}(\mathfrak{r}) = 0) \tag{3.1}$$

(read: incompatibility of $\underline{\varepsilon}(\mathfrak{r}) = 0$) then the (symmetric) incompatibility tensor $\underline{\eta}(\mathfrak{r})$ in the extended equations

$$\text{Ink}\ \underline{\varepsilon}^{el}(\mathfrak{r}) = \underline{\eta}(\mathfrak{r}) \tag{3.2}$$

is a measure of the self-stresses, i.e. it takes over the role of a source function for self-stresses.

A macroscopically homogeneous distribution of point-defects

---

[1] Vectors and tensors of the 3-dimensional physical (Euclidean) space E are written in component (subscript) notation or in symbolic form. The first notation refers to cartesian coordinates. The symbolic form uses gothic letters for vectors, simply underlined latin or greek letters for tensors of 2nd rank and doubly underlined latin or greek letters for tensors of 4th rank.

leads to a macroscopically stress-free strain which is similar to that produced by an increase of temperature. It is then plausible that the strength and concentration of these defects around point $\mathfrak{r}$, say, can be described by an external or extra strain $\underline{\varepsilon}^{ex}(\mathfrak{r})$ (see RIEDER /2/). The total strain $\underline{\varepsilon}^{tot}(\mathfrak{r})$ which is always compatible is composed of the elastic and extra strain:

$$\underline{\varepsilon}^{tot}(\mathfrak{r}) = \underline{\varepsilon}^{el}(\mathfrak{r}) + \underline{\varepsilon}^{ex}(\mathfrak{r}) \quad . \tag{3.3}$$

It follows that the incompatibility tensor which describes a distribution of point defects can be obtained by ordinary differentiations in the form

$$\underline{\eta}(\mathfrak{r}) = - \operatorname{Ink} \underline{\varepsilon}^{ex}(\mathfrak{r}) \quad . \tag{3.4}$$

In the continuum theory of dislocations a distribution of dislocations is described by an asymmetric tensor field $\underline{\alpha}(\mathfrak{r})$ from which the incompatibility field $\underline{\eta}(\mathfrak{r})$ can be derived as

$$\underline{\eta}(\mathfrak{r}) = (\underline{\alpha}(\mathfrak{r}) \times \nabla)_{sym} \tag{3.5}$$

(see KRÖNER /3/). We have denoted by "sym" the symmetrization of the tensor in parentheses. Altogether we have found that the self-stress problem of the point-defects and dislocations can be comprised in the incompatibility problem. This problem consists in the task of solving the eqs.(3.2) with given $\underline{\eta}(\mathfrak{r})$ together with the equilibrium and boundary conditions.

We now consider the following two standard boundary value problems only:

1st boundary value problem

$$\operatorname{Ink} \underline{\varepsilon}(\mathfrak{r}) = \underline{\eta}(\mathfrak{r}),\ \operatorname{Div} \underline{\sigma}(\mathfrak{r}) = - \mathfrak{f}(\mathfrak{r}) \text{ in } V; \mathfrak{s}(\mathfrak{r}) = \bar{\mathfrak{s}}(\mathfrak{r}) \text{ on } S$$

2nd boundary value problem

$$\operatorname{Ink} \underline{\varepsilon}(\mathfrak{r}) = \underline{\eta}(\mathfrak{r}),\ \operatorname{Div} \underline{\sigma}(\mathfrak{r}) = - \mathfrak{f}(\mathfrak{r}) \text{ in } V; \mathfrak{n}\cdot\underline{\sigma}(\mathfrak{r}) = \mathfrak{A}(\mathfrak{r}) \text{ on } S.$$

Here we have denoted by $\underline{\sigma}(\mathfrak{r})$ the stress tensor at point $\mathfrak{r}$, by $\mathfrak{f}(\mathfrak{r})$ the density of volume forces, by $\mathfrak{A}(\mathfrak{r})$ the density of surface forces, by $\mathfrak{s}(\mathfrak{r})$ the displacements in the volume V, by $\bar{\mathfrak{s}}(\mathfrak{r})$ the displacements on the surface S, and by $\mathfrak{n}$ the outward unit vector normal to the surface.

As we shall see, the incompatibility problem ($\underline{\eta}$ given) can be solved once the problem with $\underline{\eta} = 0$ is solved.

4. The Method of the Modified Green's Functions

We assume that the material obey a linear law of elasticity of the form

$$\sigma_{ij}(\mathfrak{r}) = c_{ijkl}(\mathfrak{r})\varepsilon_{kl}(\mathfrak{r}) \qquad (4.1)$$

where we have admitted an inhomogeneous distribution of the elastic moduli. Introducing the (volume and surface) differential operators

$$D_{ik}(\nabla) \equiv \partial_j c_{ijkl}(\mathfrak{r})\partial_l \;,\; N_{ik}(\nabla) \equiv n_j c_{ijkl}(\mathfrak{r})\partial_l \qquad (4.2)$$

we find the well-known formula of SOMIGLIANA /4/ and FREDHOLM /5/

$$s_m(\mathfrak{r}) = -\int_V dV' G_{mk}(\mathfrak{r},\mathfrak{r}')D_{kl}(\nabla')s_l(\mathfrak{r}') + \int dS' G_{mk}(\mathfrak{r},\mathfrak{r}')N_{kl}(\nabla')s_l(\mathfrak{r}')$$
$$- \int_S dS' s_i(\nabla')N_{ik}(\mathfrak{r}')G_{mk}(\mathfrak{r},\mathfrak{r}') \qquad (4.3)$$

in which $s_m(\mathfrak{r})$ are the components of the displacement field $\mathfrak{s}(\mathfrak{r})$. The formula is valid if the tensor function[1]

$$G_{mk}(\mathfrak{r},\mathfrak{r}') = G_{km}(\mathfrak{r}',\mathfrak{r}) \qquad (4.4)$$

satisfies the differential equation

$$D_{ik}(\nabla')G_{km}(\mathfrak{r},\mathfrak{r}') + \delta_{im}\delta(\mathfrak{r},\mathfrak{r}') = 0 \;. \qquad (4.5)$$

From (4.3) we form the elastic strains and obtain after partial integration

---

[1] Eq.(4.4) means that $G_{mk}(\mathfrak{r},\mathfrak{r}')$ is self-adjoint. It can be shown that beyond this property $G_{mk}(\mathfrak{r},\mathfrak{r}')$ is symmetric in m,k and in $\mathfrak{r},\mathfrak{r}'$. These properties are lost if non-local elasticity is included (see below) whereas eq. (4.4) remains valid even then and is sufficient for our later considerations to hold.

$$\varepsilon_{mn}(\mathbf{r})=\int dV'\Gamma_{mnij}(\mathbf{r},\mathbf{r}')\sigma_{ij}(\mathbf{r}')-\int dS'\Gamma_{mnij}(\mathbf{r},\mathbf{r}')c_{klij}(\mathbf{r}')n'_k\bar{s}_l(\mathbf{r}'). \tag{4.6}$$

We call Γ_{mnij} the components of a modified Green's (tensor) function. They are defined by

$$\Gamma_{mnij}(\mathbf{r},\mathbf{r}') = G_{mi,nj'}(\mathbf{r},\mathbf{r}')\Big|_{(mn)(ij)} . \tag{4.7}$$

Here we have used the comma notation for differentiation and (ij) means symmetrization in i,j. Because of (4.4) the tensor function $\underline{\underline{\Gamma}}(\mathbf{r},\mathbf{r}')$, too, is self-adjoint:[1]

$$\Gamma_{mnij}(\mathbf{r},\mathbf{r}') = \Gamma_{ijmn}(\mathbf{r}',\mathbf{r}) . \tag{4.8}$$

The $\underline{G}(\mathbf{r},\mathbf{r}')$ which enters eq.(4.6) via $\underline{\underline{\Gamma}}(\mathbf{r},\mathbf{r}')$ is now replaced by the Green's functions $\underline{G}^{(1)}(\mathbf{r},\mathbf{r}')$ or $\underline{G}^{(2)}(\mathbf{r},\mathbf{r}')$ of the 1st and 2nd boundary value problem. These functions satisfy, in addition to eq.(10) the boundary conditions

$$G^{(1)}_{km}(\mathbf{r},\mathbf{r}') = 0,\ N_{ik}(\nabla')G^{(2)}_{km}(\mathbf{r},\mathbf{r}')+\delta_{im}/S = 0 \text{ on } S \tag{4.9}$$

respectively. We thus obtain for the first boundary value problem

$$\varepsilon^{el}_{mn}(\mathbf{r}) = \int dV'\Gamma^{(1)}_{mnij}(\mathbf{r},\mathbf{r}')\sigma_{ij}(\mathbf{r}'),\ \text{if } \bar{s}_l = 0 \tag{4.10}$$

and for the 2nd boundary value problem

$$\varepsilon^{el}_{mn}(\mathbf{r}) = \int dV'\Gamma^{(2)}_{mnij}(\mathbf{r},\mathbf{r}')\sigma_{ij}(\mathbf{r}') . \tag{4.11}$$

Both equations hold under the condition $\underline{\eta} = 0$.

Following DEDERICHS and ZELLER /6,7/, we shall now introduce an abridged notation which is more common in the quantum mechanical field theory. In this notation we write in a very compact form all equations which contain only symmetric 1-point tensors of 2nd rank like $\underline{\varepsilon}(\mathbf{r})$ and self-adjoint 2-point tensors of 4th rank like $\underline{\underline{\Gamma}}(\mathbf{r},\mathbf{r}')$ in space E. Both kind of tensors will now be understood as quantities in a

---

[1]Note that $\Gamma_{mnij}(\mathbf{r},\mathbf{r}')$ is neither symmetric in $\mathbf{r},\mathbf{r}'$ nor in the exchange mn↔ij. This statement refers to local as well as to non-local elasticity.

Hilbert space H, namely as vectors and as tensors of 2nd rank in H respectively. Integrations as in eq.(4.10) are then inner multiplications in H. In the abridged notation we renounce all subscripts and arguments. The eqs.(4.10, 4.11) then assume the simple form

$$\underline{\varepsilon} = \underline{\underline{\Gamma}}^{(1)}\underline{\sigma} \quad (\underline{\eta} = 0,\ \bar{\mathfrak{s}} = 0) \tag{4.12}$$

$$\underline{\varepsilon} = \underline{\underline{\Gamma}}^{(2)}\underline{\sigma} \quad (\underline{\eta} = 0) \quad . \tag{4.13}$$

In order to complete our nomenclature we introduce the unit tensor $\underline{\underline{I}}$ of the combined space (E,H). It is the product of the unit tensor of 4th rank $\underline{\underline{I}}_4$ in E which has the components $(\delta_{ik}\delta_{jl} + \delta_{il}\delta_{jk})/2$ and the unit tensor of H which is the Dirac delta function $\delta(\mathfrak{r},\mathfrak{r}')$. In order to be consistent the tensor $\underline{\underline{c}}$ of elasticity, too, must be understood as a 2-point function. Its components are now written as $c_{ijkl}(\mathfrak{r}')\delta(\mathfrak{r},\mathfrak{r}')$. It is interesting to note that the whole formalism, including the later main results, does also apply if the components of $\underline{\underline{c}}$ are given in the more general form $c_{ijkl}(\mathfrak{r},\mathfrak{r}')$. This is the case of the non-local elasticity. The law of elasticity

$$\underline{\sigma} = \underline{\underline{c}}\ \underline{\varepsilon} \tag{4.14}$$

then has the meaning

$$\sigma_{ij}(\mathfrak{r}) = \int dV' c_{ijkl}(\mathfrak{r},\mathfrak{r}')\varepsilon_{kl}(\mathfrak{r}') \quad . \tag{4.15}$$

If the $c_{ijkl}(\mathfrak{r},\mathfrak{r}')$ degenerate to delta functions we are left with the ordinary local law. We now compare the eqs.(4.12, 4.13) with the inverse law of elasticity

$$\underline{\varepsilon} = \underline{\underline{s}}\ \underline{\sigma} \quad , \quad \underline{\underline{s}} \equiv \underline{\underline{c}}^{-1} \quad . \tag{4.16}$$

For this purpose we imagine an arbitrary tensor field $\underline{\sigma}$ on the right hand side of these equations. Such a field can always be decomposed into contributions $\underline{\sigma}^{\eta}$, $\underline{\sigma}^{\mathfrak{f}}$ and $\underline{\sigma}^{\bar{\mathfrak{s}}}$ or $\underline{\sigma}^{\mathfrak{A}}$ which can be ascribed to causes $\underline{\eta}$, $\mathfrak{f}$ and $\bar{\mathfrak{s}}$ or $\mathfrak{A}$. Then we see from eq.(4.2) that $\underline{\underline{\Gamma}}^{(1)}$ eliminates the parts $\underline{\sigma}^{\eta}$ and $\sigma^{\mathfrak{s}}$ and similarly from eq.(4.13) we can conclude that $\underline{\underline{\Gamma}}^{(2)}$ annihilates

$\underline{\sigma}^{\eta}$. In contrast, $\underline{\underline{s}}$ leaves the whole $\underline{\sigma}$ uneffected. It follows that $\underline{\underline{\Gamma}}^{(1)}$ and $\underline{\underline{\Gamma}}^{(2)}$ are projection operators which produce strains satisfying certain homogeneous conditions in the volume and partly on the surface. This is especially important in the application of variational principles.

We now substract successively the eqs.(4.12) and (4.13) from eq.(4.16). If we observe that the $\underline{\varepsilon}$ which stands on the left hand side of these equations always has a different meaning we obtain

$$\underline{\varepsilon} = (\underline{\underline{s}} - \underline{\underline{\Gamma}}^{(1)})\underline{\sigma} \qquad (\mathfrak{f} = 0) \tag{4.17}$$

$$\underline{\varepsilon} = (\underline{\underline{s}} - \underline{\underline{\Gamma}}^{(2)})\underline{\sigma} \qquad (\mathfrak{f} = 0,\ \mathfrak{A} = 0)\ . \tag{4.18}$$

It is convenient to exchange the role of $\underline{\sigma}$ and $\underline{\varepsilon}$ in these equations by using the law of elasticity. In this way we arrive at

$$\underline{\sigma} = \underline{\underline{\Delta}}^{(1)}\underline{\varepsilon} \qquad (\mathfrak{f} = 0) \tag{4.19}$$

$$\underline{\sigma} = \underline{\underline{\Delta}}^{(2)}\underline{\varepsilon} \qquad (\mathfrak{f} = 0,\ \mathfrak{A} = 0)\ . \tag{4.20}$$

Here we have introduced further modified Green's functions defined by

$$\underline{\underline{\Delta}} \equiv \underline{\underline{c}} - \underline{\underline{c}}\ \underline{\underline{\Gamma}}\ \underline{\underline{c}}\ ,\ \underline{\underline{\Gamma}} = \underline{\underline{s}} - \underline{\underline{s}}\ \underline{\underline{\Delta}}\ \underline{\underline{s}}\ . \tag{4.21}$$

We have found that by means of $\underline{\underline{\Delta}}^{(1)}$ and $\underline{\underline{\Delta}}^{(2)}$ we can create stress fields which satisfy certain homogeneous conditions in the volume and, partly, on the surface. Again this is important for the application of variational principles.

We shall not discuss here the variational principles because they would need more space than is available. Instead we briefly outline another way to solve the incompatibility problem, whereby we confine ourselves to the case in which $\mathfrak{f} = 0$ and $\mathfrak{A} = 0$. Let $(\nu = 1,2,3\ldots\infty)$ be a complete system of functions - for instance functions which are homogeneous in the cartesian coordinates. Then the system $\underline{\underline{\Delta}}^{(2)}\underline{\varepsilon}_{\nu}$ is no longer complete in general. It remains complete, however, with reference to all stress states which satisfy $\mathfrak{f} = 0$ and $\mathfrak{A} = 0$. Now $\underline{\eta}$ can be written as a superposition

$$\underline{\eta} = \sum_{\nu} a_{\nu} \ \text{Ink}(\underset{=}{s} \ \underset{=}{\Delta}^{(2)} \ \underline{\varepsilon}_{\nu}) \tag{4.22}$$

of functions which can be calculated when the Green's function of the 2nd boundary value problem is known. If the coefficients a_{ν} are determined in the well-known manner which perhaps requires an orthogonalization procedure, then the problem is solved. In fact, we have

$$\underline{\sigma} = \sum_{\nu} a_{\nu} \ \underset{=}{\Delta}^{(2)} \ \underline{\varepsilon}_{\nu} \ . \tag{4.23}$$

In principle, this method is very simple. In practice, it is quite involved because the Green's functions have a rather complex structure.

The situation is much better in all cases where the sources of the self-stresses are given in the form of $\underline{\varepsilon}^{ex}$. This is the case for the point-defects but not for the dislocations. It follows from eq.(4.20) that the operator $\underset{=}{\Delta}^{(2)}$ annihilates any strain field which is produced by volume or surface forces. The totality of such strain fields has the form Def $\mathfrak{w}(\mathfrak{r})$ with arbitrary vector field $\mathfrak{w}(\mathfrak{r})$. However, this is also the form of the tensor field $\underline{\varepsilon}^{tot}$ of the total strain. If we now replace by $\underline{\varepsilon}^{tot} - \underline{\varepsilon}^{ex}$ the field $\underline{\varepsilon}^{el}$ in eq.(4.20) then the part $\underline{\varepsilon}^{tot}$ is eliminated and we arrive at

$$\underline{\sigma} = - \underset{=}{\Delta}^{(2)} \underline{\varepsilon}^{ex} \ . \tag{4.24}$$

This equation solves the problem of determining the self-stresses of a source density $\underline{\varepsilon}^{ex}$ in the absence of external forces.

The solution (4.24), like (4.23), is applicable in the case of arbitrary anisotropy and inhomogeneity of the elastic constants. It can also be used in the case of temperature and magnetostrictive stresses as in other self-stress situations. Of course, one needs explicitly the Green's function of the 2nd boundary value problem which is, generally, available only in the form of series developments. Therefore, also the last method usually entails a greater mathematical effort. Eq.(4.24) is equivalent to Maysel's formula (see NOWACKI /8/), if this is written for the stress $\underline{\sigma}$ and for a stress-free surface.

As can be seen from eqs.(4.12) and (4.20) $\underline{\underline{\Gamma}}^{(1)}$ annihilates every constant tensor $\underline{\sigma}$ and $\underline{\underline{\Delta}}^{(2)}$ does the same with every constant tensor $\underline{\varepsilon}$. This means that both $\underline{\underline{\Gamma}}^{(1)}$ and $\underline{\underline{\Delta}}^{(2)}$ eliminate every constant tensor. Since such a tensor is always reproduced by multiplication with $\underline{\underline{I}}_4$, the unit tensor of E, we obtain

$$\underline{\underline{\Gamma}}^{(1)} \underline{\underline{I}}_4 = 0 , \quad \underline{\underline{\Delta}}^{(2)} \underline{\underline{I}}_4 = 0 , \tag{4.25}$$

and this is identical with

$$\int dV' \underline{\underline{\Gamma}}^{(1)}(\mathfrak{r},\mathfrak{r}') = 0, \quad \int dV' \underline{\underline{\Delta}}^{(2)}(\mathfrak{r},\mathfrak{r}') = 0 . \tag{4.26}$$

Furthermore, multiplying the eqs.(4.21) by $\underline{\underline{s}}\underline{\underline{I}}_4$ and $\underline{\underline{c}}\underline{\underline{I}}_4$ respectively from the right and observing eqs.(4.25) we find

$$\underline{\underline{\Delta}}^{(1)}\underline{\underline{s}}\underline{\underline{I}}_4 = \underline{\underline{I}}_4 , \quad \underline{\underline{\Gamma}}^{(2)}\underline{\underline{c}}\underline{\underline{I}}_4 = \underline{\underline{I}}_4 . \tag{4.27}$$

In the case of local, homogeneous $\underline{\underline{c}}$ and $\underline{\underline{s}}$ this is equivalent to

$$\int dV' \underline{\underline{\Delta}}^{(1)}(\mathfrak{r},\mathfrak{r}') = \underline{\underline{c}} , \quad \int dV' \underline{\underline{\Gamma}}^{(2)}(\mathfrak{r},\mathfrak{r}') = \underline{\underline{s}} . \tag{4.28}$$

I have written down the relations (4.25) and (4.27) which are consequences of the former equations because they play an important role in practical problems.

5. Concluding Remarks

At the beginning of this lecture I had discussed the physical origin of the self-stresses which normally remain after some unelastic treatment of a piece of material. We had seen that these stresses arise from sources on an atomic scale. As far as these sources do not exist in thermal equilibrium they are identical with those lattice defects which are produced in a huge number by the unelastic treatment. Macroscopically one observes the superposition of the self-stresses of all these sources.

If one does not deal with crystals but, for instance,

with glasses then the notion of lattice defects does not make sense. Nevertheless, it can be expected that the arrangement of neighbouring particles still reflects the conditions of cooling under which the glass state was produced. These conditions manifest themselves in the form of density fluctuations which again act as sources of self-stresses. The situation is similar to that of the point-defects which also could be regarded as density deviations.

The quantitative prediction of the temporal and spatial development of the sources as well as the calculation of the pertaining self-stresses are fundamental problems of the physics and technology of the glasses, as they were shown to be in the case of the crystalline materials. The first part of the problem is much more involved and has been solved only for very simplified models. The second part, the calculation of the self-stresses from the given sources is also mathematically complex, particularly because the problems are normally 3-dimensional. However, there do exist lucid formalisms among which the method of the Green's function possesses special importance.

This is discussed in the mathematical part of this lecture in which we also have treated the problem of load-stresses. Although everything can be calculated, in principle, by means of the usual Green's functions $\underline{G}^{(1)}$ and $\underline{G}^{(2)}$ of the 1st and 2nd boundary value problem it appears to be convenient in many cases to introduce a set of four modified Green's functions. These follow from $\underline{G}^{(1)}$ and $\underline{G}^{(2)}$ as given by eqs.(4.7) and (4.21). For convenience we have tabulated the main results in the following form:

$\underline{\varepsilon} = \underline{\underline{\Gamma}}^{(1)}\underline{\sigma}$ $(\underline{n} = 0,\ \bar{\sigma} = 0)$	$\underline{\varepsilon} = \underline{\underline{\Gamma}}^{(2)}\underline{\sigma}$ $(\underline{n} = 0)$	$\underline{\sigma} = \underline{\underline{\Delta}}^{(1)}\underline{\varepsilon}$ $(\mathfrak{f} = 0)$	$\underline{\sigma} = \underline{\underline{\Delta}}^{(2)}\underline{\varepsilon}$ $(\mathfrak{f} = 0,\ \mathfrak{A} = 0)$
$\underline{\underline{\Gamma}}^{(1)}\underline{\underline{I}}_4 = 0$	$\underline{\underline{\Gamma}}^{(2)}\underline{\underline{c}}\,\underline{\underline{I}}_4 = \underline{\underline{I}}_4$	$\underline{\underline{\Delta}}^{(1)}\underline{\underline{s}}\,\underline{\underline{I}}_4 = \underline{\underline{I}}_4$	$\underline{\underline{\Delta}}^{(2)}\underline{\underline{I}}_4 = 0$

The formulas of the first line apply under the conditions given in parentheses, and those of the last line apply in all cases. Besides, all formulas hold also if the elasticity tensor is anisotropic, inhomogeneous and/or non-local.

The short-hand notation of DEDERICHS and ZELLER leads to very compact formulas. It is particularly suitable for the iterative solution of problems since the multiple integrals appearing here become quite clear. To give an example: The (local or non-local) effective tensor of elasticity $\underline{\underline{C}}^{eff}$ of a random elastic medium is expressed in terms of the microscopic tensor $\underline{\underline{c}}$ by the formula (see DEDERICHS and ZELLER l.c., LEWIN /9/, MAZILU /10/ and KRÖNER /11/)

$$\underline{\underline{C}}^{eff} = \langle \underline{\underline{c}} \rangle - \langle c' \Gamma^{(1)} c' \rangle + \langle c' \Gamma^{(1)} c' \Gamma^{(1)} c' \rangle - \ldots + \ldots \quad (5.1)$$

where $\underline{\underline{c}}' = \underline{\underline{c}} - \langle c \rangle$ and the brackets denote ensemble averages. Eq.(5.1) is a development involving multiple integrals. Equations which are formally similar to eq.(5.1) also appear if we consider the fluctuating self-stresses which develop during the plastic deformation of polycrystalline materials.

Finally, it is worth-while to note that non-local elasticity is not just a mathematically interesting variant of elastic behaviour which has nothing to do with real materials. In fact, non-local effective tensors of elasticity are always obtained when the distribution of the microscopic moduli shows correlations over finite distances. There is no particular difficulty in the preparation of such materials which may turn out to be suitable for special purposes.

Acknowledgement

The set of formulas shown in the last section was the result of research supported by the Deutsche Forschungsgemeinschaft. I would like to thank Dr. B.K. Datta-Gairola for giving me his opinion on the manuscript.

References

1. REISSNER, H.: Eigenspannungen und Eigenspannungsquellen. Z. Angew. Math. Mech. 11, 1-8 (1931).
2. RIEDER, G.: Spannungen und Dehnungen im gestörten elastischen Medium. Z. Naturforsch. 11a, 171-173 (1956).
3. KRÖNER, E.: Kontinuumstheorie der Versetzungen und Eigenspannungen. Erg. Angew. Math. 5, 1-179 (1958).
4. SOMIGLIANA, C.: Sulle equazioni della elasticita. Ann. Mat. (2) 17, 37-64 (1889).
5. FREDHOLM, I.: Sur les équations de l'équilibre d'un corps solide élastique. Acta math. 23, 1-42 (1900).
6. ZELLER, R. and P.H. DEDERICHS: Elastic constants of polycrystals. phys. stat. sol. (b) 55, 831-842 (1973).
7. DEDERICHS, P.H. and R. ZELLER: Variational treatment of the elastic constants of disordered materials. Z. Physik 259, 103-116 (1973).
8. NOWACKI, W.: Thermoelasticity, Oxford, Pergamon Press 1962.
9. LEVIN, V.M.: The relation between mathematical expectations of stress and strain tensors in elastic microheterogeneous media. Prikl. Mat. Mech. 35, 694-701 (1971), transl. from Russ.
10. MAZILU, P.: On the theory of linear elasticity in statistically homogeneous media. Rev. Roum. Mat. Pur. Appl. 17, 261-273 (1972).
11. KRÖNER, E.: Zur klassischen Theorie statistisch aufgebauter Festkörper. Int. J. Engng. Sci. 11, 171-191 (1973).

DISTORTION IN MICROPOLAR ELASTICITY

W. NOWACKI, Warsaw

1. Fundamental relations and equations

Consider a micropolar, isotropic, homogeneous and centro-symmetric elastic body, subject to initial strain $\gamma_{ji}^{0}, \varkappa_{ji}^{0}$ depending on position $\underline{x}$. This strain can arise in metals in exceeding the yield limit or during changes occuring in a heat working. A special case of distortions is the temperature strain $\gamma_{ji}^{0} = \alpha_t \delta_{ji}\, \theta$, $\varkappa_{ji}^{0} = 0$ where α_t denotes the coefficient of linear thermal expansion and $\theta = T - T_0$ is the temperature increase. $T(\underline{x})$ is the absolute temperature at point $\underline{x}$ and $T_0 =$ const is the temperature of the natural state.

We assume that the strain $\gamma_{ji}^{0}, \varkappa_{ji}^{0}$ is of the same order as the elastic strain. The introduction of the initial strain $\gamma_{ji}^{0}, \varkappa_{ji}^{0}$ into the body produces a state of elastic strain $\gamma_{ji}', \varkappa_{ji}'$ and the state of stress and couple stress σ_{ji}, μ_{ji}. The total strain $\gamma_{ji}, \varkappa_{ji}$ consists of two parts: the initial strain $\gamma_{ji}^{0}, \varkappa_{ji}^{0}$ and the elastic strain $\gamma_{ji}', \varkappa_{ji}'$, i.e.

$$\gamma_{ji} = \gamma_{ji}^{0} + \gamma_{ji}', \quad \varkappa_{ji} = \varkappa_{ji}^{0} + \varkappa_{ji}' \tag{1.1}$$

The elastic strain $\gamma_{ji}', \varkappa_{ji}'$ is a linear function of the stress [1]

$$\begin{aligned} \gamma_{ji}' &= (\mu' + \alpha')\, \sigma_{ji} + (\mu' - \alpha')\, \sigma_{ij} + \lambda' \delta_{ij} \sigma_{kk} \,, \\ \varkappa_{ji}' &= (\gamma' + \epsilon')\, \mu_{ji} + (\gamma' - \epsilon')\, \mu_{ij} + \beta' \delta_{ij} \mu_{kk} \,, \end{aligned} \tag{1.2}$$

where $\mu', \lambda', \alpha', \beta', \gamma', \epsilon'$ are material constants.

Substituting (1.2) into (1.1) and solving the latter equations for stresses we obtain

$$\begin{aligned} \sigma_{ji} &= (\mu + \alpha)(\gamma_{ji} - \gamma_{ji}^{0}) + (\mu - \alpha)(\gamma_{ij} - \gamma_{ij}^{0}) + \lambda \delta_{ij} (\gamma_{kk} - \gamma_{kk}^{0}) \\ \mu_{ji} &= (\gamma + \epsilon)(\varkappa_{ji} - \varkappa_{ji}^{0}) + (\gamma - \epsilon)(\varkappa_{ij} - \varkappa_{ij}^{0}) + \beta \delta_{ij} (\varkappa_{kk} - \varkappa_{kk}^{0}) \end{aligned} \tag{1.3}$$

where $\mu, \lambda, \alpha, \beta, \gamma, \epsilon$ are the material constants of the Cosserat medium and

$$2\mu' = \frac{1}{2\mu}\,,\quad 2\alpha' = \frac{1}{2\alpha}\,,\quad 2\gamma' = \frac{1}{2\gamma}\,,\quad 2\epsilon' = \frac{1}{2\epsilon}\,,$$

$$\lambda' = -\frac{\lambda}{2\mu(3\lambda+2\mu)}\,,\quad \beta' = -\frac{\beta}{2\gamma(3\beta+2\gamma)}\,.$$

The total strain $\gamma_{ji}, \varkappa_{ji}$ can be expressed in terms of the displacement vector $\underline{u}$ and the rotation vector $\underline{\varphi}$ as follows [1]:

$$\gamma_{ji} = u_{i,j} - \epsilon_{kji}\varphi_k\,,\quad \varkappa_{ji} = \varphi_{i,j} \tag{1.4}$$

If the stress σ_{ji}, μ_{ji} from the formulae (1.3) is introduced into the equilibrium equations

$$\sigma_{ji,j} = 0\,,\quad \epsilon_{ijk}\sigma_{jk} + \mu_{ji,j} = 0 \tag{1.5}$$

and the relations (1.4) are taken into account, then we arrive at a system of six equations in displacements and rotations

$$\left\{\begin{array}{l} (\mu+\alpha)\nabla^2\underline{u} + (\lambda+\mu-\alpha)\,\mathrm{grad\,div}\,\underline{u} + 2\alpha\,\mathrm{rot}\,\underline{\varphi} + \underline{X} = 0\,, \\ \\ ((\gamma+\epsilon)\,\nabla^2 - 4\alpha)\,\underline{\varphi} + (\beta+\gamma-\epsilon)\,\mathrm{grad\,div}\,\underline{\varphi} + 2\alpha\,\mathrm{rot}\,\underline{u} + \underline{Y} = 0\,, \end{array}\right. \tag{1.6}$$

We have introduced here fictitious body forces

$$X_j = -\overset{0}{\sigma}_{ji,j}\,,\quad Y_i = -\epsilon_{ijk}\overset{0}{\sigma}_{jk} - \overset{0}{\mu}_{ji,j}\,, \tag{1.7}$$

where

$$\begin{aligned} \overset{0}{\sigma}_{ji} &= (\mu+\alpha)\,\overset{0}{\gamma}_{ji} + (\mu-\alpha)\,\overset{0}{\gamma}_{ij} + \lambda\delta_{ij}\overset{0}{\gamma}_{kk}\,, \\ \overset{0}{\mu}_{ji} &= (\gamma+\epsilon)\,\overset{0}{\varkappa}_{ji} + (\gamma-\epsilon)\,\overset{0}{\varkappa}_{ij} + \beta\delta_{ij}\overset{0}{\varkappa}_{kk}\,. \end{aligned} \tag{1.8}$$

Eqs (1.6) should be completed by boundary conditions which may be given in displacements and rotations or in surface forces and moments on the surface A bounding the body.

Solving the differential equations (1.6) we obtain displacement $\underline{u}$ and rotation $\underline{\varphi}$. Eqs (1.4) serve then for the determination of the strain $\gamma_{ji}, \varkappa_{ji}$ and (1.3) makes it possible to calculate the stress σ_{ji}, μ_{ji}.

Eqs (1.6) are particularly simple if we are faced with thermal distortions, namely

$$\overset{0}{\gamma}_{ji} = \alpha_t\delta_{ij}\theta(\underline{x})\,,\quad \overset{0}{\varkappa}_{ji} = 0 \tag{1.9}$$

Then they have the form

$$\begin{cases} (\mu+\alpha)\,\nabla^2\underline{u}+(\lambda+\mu-\alpha)\,\mathrm{grad\,div}\,\underline{u}+2\alpha\,\mathrm{rot}\,\varphi=\nu\,\mathrm{grad}\,\theta\ , \\ ((\gamma+\epsilon)\,\nabla^2-4\alpha)\,\underline{\varphi}+(\beta+\gamma-\epsilon)\,\mathrm{grad\,div}\,\underline{\varphi}+2\alpha\,\mathrm{rot}\,\underline{u}=0\,,\ \nu=(3\lambda+2\mu)\,\alpha_t\ . \end{cases} \tag{1.10}$$

A different method of determination of the state of stress due to the action of a distortion was given by K.H. Anthony [2] and W.D.Claus and A.C.Eringen [3].

In view of relations (1.1) and (1.4) we have

$$u_{i,j}-\epsilon_{kji}\varphi_k=\overset{0}{\gamma}_{ji}+\gamma'_{ji}\,,\quad \varphi_{i,j}=\overset{0}{\varkappa}_{ji}+\varkappa'_{ji}\ . \tag{1.11}$$

Eliminating from Eqs (1.11) the quantities φ_i and u_i we arrive at the compatibility equations

$$\begin{cases} \epsilon_{jhl}\gamma'_{li,h}-\varkappa'_{ij}+\delta_{ij}\varkappa'_{kk}=\alpha_{ji}\ , \\ \epsilon_{jhl}\varkappa'_{li,h}=\theta_{ji}\ , \end{cases} \tag{1.12}$$

where

$$\begin{cases} \alpha_{ji}=-\epsilon_{jhl}\overset{0}{\gamma}_{li,h}+\overset{0}{\varkappa}_{ij}-\delta_{ij}\overset{0}{\varkappa}_{kk}\ , \\ \theta_{ji}=-\epsilon_{jhl}\overset{0}{\varkappa}_{li,h}\ . \end{cases}$$

The quantities α_{ji} and θ_{ji} are known functions and constitute the distortion densities. Representing the elastic strain γ'_{ji}, $\varkappa'_{ji}$ in terms of the stress σ_{ji}, μ_{ji} by means of relations (1.9) we obtain the compatibility equations in stresses. Making use of the equations of equilibrium (1.5) we arrive at equations in stresses constituting the counterpart of the Beltrami-Michell equations of classical elasticity. Eqs (1.12) are particularly convenient in the case of plane state of strain.

In the particular case of thermal distortion $\overset{0}{\gamma}_{ji}=\alpha_t\theta\,\delta_{ji}$, $\overset{0}{\varkappa}_{ji}=0$ we have $\theta_{ji}=0$, $\alpha_{ji}=\epsilon_{jih}\theta_{,h}$ and the compatibility equations take the form

$$\begin{cases} \epsilon_{jhl}\gamma'_{li,h}-\varkappa'_{ij}+\delta_{ij}\varkappa'_{kk}=\alpha_t\epsilon_{jih}\theta_{,h}\ , \\ \epsilon_{jhl}\varkappa'_{li,h}=0\ . \end{cases} \tag{1.13}$$

2. Principle of virtual work

Consider a micropolar body in equilibrium, subject to external loading (body forces and moments $\underline{X}, \underline{Y}$ and surface forces and moments $\underline{p}, \underline{m}$) and a field of initial strain $\overset{0}{\gamma}_{ji}, \overset{0}{\varkappa}_{ji}$. Suppose that on the surface A_σ forces and moments are given, while on A_u displacements and rotations are prescribed. We have $A = A_u + A_\sigma$ where A is the total surface bounding the body.

The principle of virtual work for virtual displacement δu_i and virtual rotation $\delta\varphi_i$ has the form

$$\int_V (X_i \delta u_i + Y_i \delta\varphi_i)\, dV + \int_A (p_i \delta u_i + m_i \delta\varphi_i)\, dA = \int_V (\sigma_{ji}\delta\gamma_{ji} + \mu_{ji}\delta\varkappa_{ji})\, dV \; . \tag{2.1}$$

This principle states that the virtual work of the external forces and moments is equal to the virtual work of the internal forces.

Introducing into (2.1) the constitutive relations (1.3) we obtain the equation

$$\int_A (X_i \delta u_i + Y_i \delta\varphi_i)\, dV + \int_A (p_i \delta u_i + m_i \delta\varphi_i)\, dA = dW - \int_V (\overset{0}{\sigma}_{ji}\, \delta\gamma_{ji} + \overset{0}{\mu}_{ji}\delta\varkappa_{ji})\, dV \; , \tag{2.2}$$

where

$$W = \int_V (\mu\gamma_{(ij)}\gamma_{(ij)} + \alpha\gamma_{<ij>}\gamma_{<ij>} + \frac{\lambda}{2}\gamma_{kk}\gamma_{nn} + \gamma\varkappa_{(ij)}\varkappa_{(ij)} + \epsilon\varkappa_{<ij>}\varkappa_{<ij>} + \frac{\beta}{2}\varkappa_{kk}\varkappa_{nn})\, dV \; .$$

The symbols ( ) and <> refer to the symmetric and skew-symmetric parts of the tensor, respectively. Since the body forces and moments and the surface forces and moments are not subject to any variation, (2.2) can be written in the form

$$\delta\,[W - \int_V (X_i u_i + Y_i \varphi_i)\, dV - \int_{A_\sigma} (p_i u_i - m_i \varphi_i)\, dA - \int_V (\overset{0}{\gamma}_{ji}\sigma_{ji} + \overset{0}{\varkappa}_{ji}\mu_{ji})\, dV] = 0 \, . \tag{2.3}$$

We have made use here of the identity

$$\overset{0}{\sigma}_{ji}\gamma_{ji} + \overset{0}{\mu}_{ji}\varkappa_{ji} = \sigma_{ji}\overset{0}{\gamma}_{ji} + \mu_{ji}\overset{0}{\varkappa}_{ji} \; . \tag{2.4}$$

The expression in the square brackets in Eq.(2.3) is the potential energy. This energy takes an extremum value. A procedure analogous to that in classical elasticity proves that the potential energy takes the absolute minimum.

Let us now return to Eq.(2.2). Making use of relations (1.4) and the Ostrogradski-Gauss theorem we obtain

$$\int_V [(X_i - \overset{0}{\sigma}_{ji,j})\, \delta u_i + (Y_i - (\overset{0}{\mu}_{ji,j} + \epsilon_{ijk}\overset{0}{\sigma}_{jk}))\, \delta\varphi_i]\, dV$$

$$+ \int_A [(p_i + \overset{0}{\sigma}_{ji} n_j)\delta u_i + (m_i + \overset{0}{\mu}_{ji} n_j)\, \delta\varphi_i]\, dA = \delta W \, . \tag{2.5}$$

Consider now the same bounded body, of the same shape and the same material. Assume that it is subject to the external body forces and moments X_i^*, Y_i^*; on A_σ there act forces p_i^* and moments m_i^* and on A_u displacements u_i^* and rotations φ_i^* are prescribed. We assume however that initial strain is absent.

We now ask the question, what should the quantities X_i^* and Y_i^* in the interior of the body be and what quantities p_i^*, m_i^* on A_σ with the same boundary conditions on A_u should be prescribed, in order that in the considered body there occur the same fields of displacement $\underline{u}$ and rotation $\underline{\varphi}$ as in the case of action of distortion $\overset{0}{\gamma}_{ji}$, $\overset{0}{\varkappa}_{ji}$. To answer it we write down the equation of virtual work for the considered body

$$\int_V (X_i^*\delta u_i + Y_i^*\delta\varphi_i)\, dV + \int_A (p_i^*\delta u_i + m_i^*\delta\varphi_i)\, dA = \delta W \;. \tag{2.6}$$

Since displacement $\underline{u}$ and rotation $\underline{\varphi}$ are the same in both cases, the right-hand sides of Eqs (2.5) and (2.6) are identical. Comparing the left-hand sides of (2.5) and (2.6) we obtain the relations

$$\begin{cases} X_i^* = X_i - \overset{0}{\sigma}_{ji,j}\,, \quad Y_i^* = Y_i - (\epsilon_{ijk}\overset{0}{\sigma}_{jk} + \overset{0}{\mu}_{ji,j})\,, & \underline{x} \epsilon V\,, \\ p_i^* = p_i + \overset{0}{\sigma}_{ji} n_j\,, \quad m_i^* = m_i + \overset{0}{\mu}_{ji} n_j\,, & \underline{x} \epsilon A_\sigma\,, \\ u_i^* = u_i\,, \qquad \varphi_i^* = \varphi_i\,, & \underline{x} \epsilon A_u\;. \end{cases} \tag{2.7}$$

The above quantities represent the counterparts of the body forces and moments. This analogy makes it possible to reduce every static distortion problem to a boundary problem of the non-symmetric elasticity theory.

3. Theorem of minimum of the complementary work

Consider the quadratic form

$$W'_\sigma = \left(\frac{\mu'+\alpha'}{2}\right)\sigma_{ji}\sigma_{ji} + \left(\frac{\mu'-\alpha'}{2}\right)\sigma_{ij}\sigma_{ji} + \frac{\lambda'}{2}\sigma_{kk}\sigma_{nn} + \left(\frac{\gamma'+\epsilon'}{2}\right)\mu_{ji}\mu_{ji} + \left(\frac{\gamma'-\epsilon'}{2}\right)\mu_{ji}\mu_{ij} + \frac{\beta}{2}\mu_{kk}\mu_{nn}\;. \tag{3.1}$$

In view of (1.1) and (1.2), we obtain

$$\frac{\partial W'_\sigma}{\partial \sigma_{ji}} = (\mu'+\alpha')\,\sigma_{ji} + (\mu'-\alpha')\,\sigma_{ij} + \lambda'\sigma_{kk}\,\delta_{ij} = \gamma_{ji} - \overset{0}{\gamma}_{ji}\,,$$

$$\frac{\partial W'_\sigma}{\partial \mu_{ji}} = (\gamma'+\epsilon')\,\mu_{ji} + (\gamma'-\epsilon')\,\mu_{ij} + \beta'\mu_{kk}\,\delta_{ij} = \varkappa_{ji} - \overset{0}{\varkappa}_{ji} \tag{3.2}$$

where we have introduced the virtual increments of stress $\delta\sigma_{ji}$, $\delta\mu_{ji}$. We assume that the virtual stress satisfies the equilibrium equations

$$\delta\sigma_{ji,j}=0\,,\quad \epsilon_{ijk}\,\delta\sigma_{jk}+\delta\mu_{ji,j}=0\,, \tag{3.3}$$

and that on surface A_σ we have $\delta p_i=0$, $\delta m_i=0$. On surface $A_u=A-A_\sigma$ the virtual increments δp_i, δm_i are arbitrary. Let us multiply relation $(3.2)_1$ by $\delta\sigma_{ji}$ and $(3.2)_2$ by $\delta\mu_{ji}$. Adding the results and integrating over the volume of the body we obtain

$$\int_V \left(\frac{\partial W'_\sigma}{\partial\sigma_{ji}}\delta\sigma_{ji}+\frac{\partial W'_\sigma}{\partial\mu_{ji}}\delta\mu_{ji}\right)dV=\int_V\left[(\gamma_{ji}-\overset{0}{\gamma}_{ji})\,\delta\sigma_{ji}+(\varkappa_{ji}-\overset{0}{\varkappa}_{ji})\,\delta\mu_{ji}\right]dV \tag{3.4}$$

or

$$\delta W_\sigma+\int_V(\overset{0}{\gamma}_{ji}\delta\sigma_{ji}+\overset{0}{\varkappa}_{ji}\delta\mu_{ji})\,dV=\int_V(\gamma_{ji}\delta\sigma_{ji}+\varkappa_{ji}\delta\mu_{ji})\,dV,\quad W_\sigma=\int_V W'_\sigma\,dV \tag{3.5}$$

On the other hand, transforming the right-hand side of Eq.(3.5) we have

$$\begin{cases}\int_V(\gamma_{ji}\delta\sigma_{ji}+\varkappa_{ji}\delta\mu_{ji})\,dV=\int_A(u_i\delta p_i+\varphi_i\delta m_i)\,dV\\[2mm] -\int_V[u_i\delta\sigma_{ji,j}+\varphi_i\,(\epsilon_{ijk}\,\delta\sigma_{jk}+\delta\mu_{ji,j})]\,dV,\quad \delta p_i=\delta\sigma_{ji}n_j\,,\quad \delta m_i=\delta\mu_{ji}n_j\,.\end{cases} \tag{3.6}$$

We have made use here of the definition of strain (1.4). Taking into account the equilibrium equation (3.3) and bearing in mind that $\delta p_i=0$, $\delta m_i=0$ on A_σ we obtain from (3.5) and (3.6)

$$\delta W_\sigma+\int_V(\overset{0}{\gamma}_{ji}\delta\sigma_{ji}+\overset{0}{\varkappa}_{ji}\delta\mu_{ji})\,dV=\int_{A_u}(u_i\delta p_i+\varphi_i\delta m_i)\,dA\,, \tag{3.7}$$

or

$$\delta\Gamma=0\,,$$

where

$$\Gamma=W_\sigma+\int_V(\overset{0}{\gamma}_{ji}\sigma_{ji}+\overset{0}{\varkappa}_{ji}\mu_{ji})\,dV-\int_{A_u}(p_iu_i+m_i\varphi_i)\,dA\,.$$

Here Γ is the complementary work. As in the classical elasticity we can prove that the complementary energy takes the absolute minimum.

4. Reciprocity theorem

In deriving this theorem we make use of the analogy of body forces and moments. The reciprocity theorem for a body without initial strain has the form

$$\int_V (X_i^* u_i' + Y_i^* \varphi_i')\, dV + \int_A (p_i^* u_i' + m_i^* \varphi_i')\, dA$$
$$= \int_V (X_i'^* u_i + Y_i'^* \varphi_i)\, dV + \int_A (p_i'^* u_i + m_i'^* \varphi_i)\, dA \tag{4.1}$$

where X_i^* , Y_i^* , p_i^* , m_i^* refer to the first system of loadings producing displacement u_i and the rotation φ_i , while $X_i'^*$, $Y_i'^*$, $p_i'^*$, $m_i'^*$ refer to the second system of loadings leading to displacement u_i' and rotation φ_i'.

Consider now the same body with the first system of loadings X_i , Y_i , p_i , m_i and the initial strain $\overset{0}{\gamma}_{ji}$, $\overset{0}{\varkappa}_{ji}$ leading to displacement and rotation fields u_i and φ_i , respectively. The second system of loadings, initial strain and rotation will be devoted by primes. Making use of the analogy of body forces and moments (2.7) we obtain

$$\int_V [(X_i - \overset{0}{\sigma}_{ji,j})\, u_i' + (Y_i - (\epsilon_{ijk} \overset{0}{\sigma}_{jk} + \overset{0}{\mu}_{ji,j}))\, \varphi_i']\, dV + \int_A [(p_i + \overset{0}{\sigma}_{ji} n_j)\, u_i' + (m_i + \overset{0}{\mu}_{ji} n_j)\, \varphi_i']\, dA$$
$$= \int_V [(X_i' - \overset{\prime 0}{\sigma}_{ji,j})\, u_i + (Y_i' - (\epsilon_{ijk} \overset{\prime 0}{\sigma}_{jk} + \overset{\prime 0}{\mu}_{ji,j}))\, \varphi_i]\, dV + \int_A [(p_i' + \overset{\prime 0}{\sigma}_{ji} n_j)\, u_i + (m_i' + \overset{\prime 0}{\mu}_{ji} n_j)\, \varphi_i]\, dA\,. \tag{4.2}$$

After simple transformations, making use of the Ostrogradski-Gauss theorem, we arrive at the final form of the generalization of the reciprocity theorem to distortion problems:

$$\int_V (X_i u_i' + Y_i \varphi_i')\, dV + \int_A (p_i u_i' + m_i \varphi_i')\, dA + \int_V (\overset{0}{\gamma}_{ji} \sigma_{ji}' + \overset{0}{\varkappa}_{ji} \mu_{ji}')\, dV$$
$$= \int_V (X_i' u_i + Y_i' \varphi_i)\, dV + \int_A (p_i' u_i + m_i' \varphi_i)\, dA + \int_V (\overset{\prime 0}{\gamma}_{ji} \sigma_{ji} + \overset{\prime 0}{\varkappa}_{ji} \mu_{ji})\, dV\,. \tag{4.3}$$

Consider a bounded body, free of body forces ($X_i = 0$) and body moments ($Y_i = 0$) , free of loadings on A_σ , i.e. $p_i = 0$, $m_i = 0$ and clamped on A_u, $\underline{u} = 0$, $\underline{\varphi} = 0$. Our aim is to determine displacement u_i and rotation φ_i at a point $\underline{\xi}$ of the body, due to the action of initial strain $\overset{0}{\gamma}_{ji}$, $\overset{0}{\varkappa}_{ji}$.

We take for the second system of loadings a concentrated force $X_i' = \delta(\underline{x} - \underline{\xi})\, \delta_{ik}$ at point $\underline{\xi}$ acting in the direction of the x_k-axis. Thus, $Y_i' = 0$, $\overset{\prime 0}{\gamma}_{ji} = 0$, $\overset{\prime 0}{\varkappa}_{ji} = 0$. Moreover, we assume that u_i', φ_i' vanish on A_u and p_i', m_i' vanish on A_σ . The action of the concentrated force $X_i' = \delta(\underline{x} - \underline{\xi})\, \delta_{ik}$ produces in the body displacements $U_i^{(k)}(\underline{x}, \underline{\xi})$ and rotations $\Phi_i^{(k)}(\underline{x}, \underline{\xi})$. By means of the latter functions we determine the strain $\gamma_{ji}^{(k)}$, $\varkappa_{ji}^{(k)}$ and the stress $\sigma_{ji}^{(k)}$, $\mu_{ji}^{(k)}$, making use of the formulae

$$\sigma_{ji}^{(k)} = (\mu + \alpha)\, \gamma_{ji}^{(k)} + (\mu - \alpha)\, \gamma_{ij}^{(k)} + \lambda \delta_{ij} \gamma_{nn}^{(k)}\,,$$
$$\mu_{ji}^{(k)} = (\gamma + \epsilon)\, \varkappa_{ji}^{(k)} + (\gamma - \epsilon)\, \varkappa_{ij}^{(k)} + \beta \delta_{ij} \varkappa_{nn}^{(k)}\,. \tag{4.4}$$

Applying to the two systems of loadings the reciprocity theorem (4.3) we obtain

$$\int_V (\overset{0}{\gamma}_{ji}\sigma_{ji}^{(k)} + \overset{0}{\varkappa}_{ji}\mu_{ji}^{(k)})\, dV = \int_V \delta\,(\underline{x}-\underline{\xi})\,\delta_{ik}\,u_i(\underline{x})\, dV(\underline{x})\,,$$

whence

$$u_k(\underline{\xi}) = \int_V [\overset{0}{\gamma}_{ji}(\underline{x})\,\sigma_{ji}^{(k)}(\underline{x},\underline{\xi}) + \overset{0}{\varkappa}_{ji}(\underline{x})\,\mu_{ji}^{(k)}(\underline{x},\underline{\xi})]\, dV(\underline{x})\,. \tag{4.5}$$

We take for the second system of loadings (primed) a concentrated body moment $Y_i' = \delta\,(\underline{x}-\underline{\xi})\,\delta_{ik}$. We assume also that $X_i' = 0$, $\gamma_{ji}'^{0} = \varkappa_{ji}'^{0} = 0$ inside the body and $u_i' = \varphi_i' = 0$ on A_u; furthermore, $p_i' = 0$, $m_i' = 0$ on A_σ.

The body moment $Y_i' = \delta\,(\underline{x}-\underline{\xi})\,\delta_{ik}$ produces in the body displacement $u_i' = \hat{U}_i^{(k)}(\underline{x},\underline{\xi})$ and rotation $\varphi_i' = \hat{\Phi}^{(k)}(\underline{x},\underline{\xi})$. By means of the formulae (4.4) we determine the stress $\hat{\sigma}_{ji}^{(k)}(\underline{x},\underline{\xi})$ and $\hat{\mu}_{ji}^{(k)}(\underline{x},\underline{\xi})$. Applying to the considered states the reciprocity theorem (4.3) we obtain

$$\varphi_k(\underline{\xi}) = \int_V [\overset{0}{\gamma}_{ji}(\underline{x})\,\hat{\sigma}_{ji}^{(k)}(\underline{x},\underline{\xi}) + \overset{0}{\varkappa}_{ji}(\underline{x})\,\hat{\mu}_{ji}^{(k)}(\underline{x},\underline{\xi})]\, dV(\underline{x})\,. \tag{4.6}$$

Relations (4.5) and (4.6) constitute a generalization of V.M.Maysel's relations [4], to the distortion problem in micropolar elasticity. They also hold for an infinite elastic space. In this particular case the singular solutions $U_i^{(k)}$, $\Phi_i^{(k)}$, $\hat{U}_i^{(k)}$, $\hat{\Phi}_i^{(k)}$ are known, namely [5]

$$\left\{\begin{aligned} U_i^{(k)} &= \frac{1}{8\pi\mu}\left(\delta_{ik}\nabla^2 R - \frac{\lambda+\mu}{\lambda+2\mu}R_{,ik}\right) + B\left(l^2\left(\frac{e^{-R/l}-1}{R}\right)_{,ik} - \delta_{ik}\frac{e^{-R/l}}{R}\right)\,,\\ \Phi_i^{(k)} &= -\frac{1}{8\pi\mu}\epsilon_{kij}\left(\frac{e^{-R/l}-1}{R}\right)_{,j}\,, \end{aligned}\right. \tag{4.7}$$

and

$$\left\{\begin{aligned} \hat{U}_i^{(k)} &= -\frac{1}{8\pi\mu}\epsilon_{kij}\left(\frac{e^{-R/l}-1}{R}\right)_{,j}\,,\\ \hat{\Phi}_i^{(k)} &= \frac{1}{16\pi\mu}\left(\frac{1-e^{-R/l}}{R}\right) + \frac{1}{16\pi\alpha}\left(\frac{e^{-R/\nu}-e^{-R/l}}{R}\right)_{,ik} + \frac{\mu+\alpha}{16\pi\alpha\mu l^2}\delta_{ik}\frac{e^{-R/l}}{R} \end{aligned}\right. \tag{4.8}$$

where

$$R = (\underline{x}-\underline{\xi})\,,\quad \nu^2 = \frac{\beta+2\gamma}{4\alpha}\,,\quad l^2 = \frac{(\mu+\alpha)(\gamma+\epsilon)}{4\alpha\mu}\,,\quad B = \frac{\alpha}{4\pi\mu(\mu+\alpha)}\,.$$

Let us now consider a simple example. We shall calculate the change of volume of a bounded simply-connected body, due to the action of distortion $\overset{0}{\gamma}_{ji}$, $\overset{0}{\varkappa}_{ji}$. We assume that inside the body $X_i = 0$, $Y_i = 0$ and $p_i = 0$, $m_i = 0$ on the surface A.

The second system of loadings acting on the body consists only of the surface forces $p_i' = 1 n_i$. We are faced here with a multi-axial extension $\sigma_{ji}' = 1\,\delta_{ji}$, for only in this case

$p_i' = \sigma_{ji}' n_j = 1\, n_i$. Since $X_i' = 0$, $Y_i' = 0$, $m_i' = 0$, $\gamma_{ji}'^{0} = 0$, $\varkappa_{ji}'^{0} = 0$, Eq. (4.3) takes the form

$$\int\limits_A p_i' u_i \, dA = \int\limits_V \overset{0}{\gamma}_{ji} \sigma_{ji}' \, dV \ . \tag{4.9}$$

Now we have

$$\int\limits_A p_i' u_i \, dA = \int\limits_A \sigma_{ji}' n_j u_i \, dA = \int\limits_A u_i n_i \, dA = \Delta V$$

where ΔV is the volume increment of the body. Therefore, (4.9) yields

$$\Delta V = \int\limits_V \overset{0}{\gamma}_{kk} \, dV \ . \tag{4.10}$$

The volume increment ΔV has the form of a very simple integral formula. The constitutive relations (1.6) imply the relation

$$\sigma_{kk} = (3\lambda - 2\mu)\,(\gamma_{kk} - \overset{0}{\gamma}_{kk}) \ . \tag{4.11}$$

Eliminating $\overset{0}{\gamma}_{kk}$ from (4.10) and (4.11) and bearing in mind that

$$\Delta V = \int\limits_V u_{k,k} \, dV = \int\limits_A u_i n_i \, dA$$

we arrive at the interesting relation

$$\int\limits_V \sigma_{kk} \, dV = 0 \ . \tag{4.12}$$

Observe that the volume increment ΔV is independent of the material constants and that the integrand contains the sum of normal distortions. In the particular case of thermal distortion we have

$$\Delta V = 3\alpha_t \int\limits_V \theta \, dV \, , \quad \int\limits_V \sigma_{kk} \, dV = 0 \ . \tag{4.13}$$

5. Equations in displacements and rotations

Consider the differential equations in displacements and rotations

$$\begin{aligned} &(\mu+\alpha)\,\nabla^2 \underline{u} + (\lambda+\mu-\alpha)\,\text{grad div}\,\underline{u} + 2\alpha\,\text{rot}\,\underline{\varphi} + \underline{X} = 0\,, \\ &((\gamma+\epsilon)\,\nabla^2 - 4\alpha)\,\underline{\varphi} + (\beta+\gamma-\epsilon)\,\text{grad div}\,\underline{\varphi} + 2\alpha\,\text{rot}\,\underline{u} + \underline{Y} = 0 \ , \end{aligned} \tag{5.1}$$

where

$$X_i = -\sigma_{ji,j}\,,\quad Y_i = -\epsilon_{ijk}\overset{0}{\sigma}_{jk} - \overset{0}{\mu}_{ji,j}\,.$$

Applying in Eqs (5.1) the divergence operator we have

$$\nabla^2 \operatorname{div} \underline{u} = -\frac{1}{\lambda+2\mu}\operatorname{div}\underline{X}$$

$$H \operatorname{div}\underline{\varphi} = -\frac{1}{\beta+2\gamma}\operatorname{div}\underline{Y} \tag{5.2}$$

where

$$H = \nabla^2 - \frac{1}{\nu^2}\,,\quad \nu^2 = \frac{\beta+2\gamma}{4\alpha}\,.$$

Observe that for $\underline{X}=0$ the dilatation $\operatorname{div}\underline{u}$ is a harmonic function and for $\underline{Y}=0$ the function $\underline{\varphi}$ satisfies the homogeneous Helmholtz equation.

Applying in Eqs (5.1) the rotation operator we obtain the relations

$$\left\{\begin{aligned} &(\mu+\alpha)\,\nabla^2 \operatorname{rot} u + 2\alpha \operatorname{rot}\operatorname{rot}\underline{\varphi} = -\operatorname{rot}\underline{X}\,,\\ &((\gamma+\epsilon)\,\nabla^2 - 4\alpha)\,\underline{\varphi} + 2\alpha\operatorname{rot}\operatorname{rot}\underline{u} = -\operatorname{rot}\underline{Y}\,. \end{aligned}\right. \tag{5.3}$$

Next we apply to Eq. $(5.1)_1$ the operator $\nabla^2 D$ where $D=(\gamma+\epsilon)\nabla^2 - 4\alpha$ and making use of the relations $(5.2)_1$ and $(5.3)_2$, after transformations we arrive at the following equation containing only the displacement vector $\underline{u}$:

$$\left\{\begin{aligned} &D\nabla^2\nabla^2\underline{u} = \frac{1}{4\alpha\mu l^2}\,[2\alpha\nabla^2\operatorname{rot}\underline{Y} - (\gamma+\epsilon)\,\nabla^2 G\underline{X}\\ &-\frac{4\alpha^2}{\lambda+2\mu}\operatorname{grad}\operatorname{div}\underline{X} + \frac{(\lambda+\mu-\alpha)(\gamma+\epsilon)}{\lambda+2\mu}\operatorname{grad}\operatorname{div}G\underline{X}]\,. \end{aligned}\right. \tag{5.4}$$

We have introduced here the notations

$$D = \nabla^2 - \frac{1}{l^2}\,,\quad G = \nabla^2 - \frac{1}{\varkappa^2}\,,\quad l^2 = \frac{(\gamma+\epsilon)(\mu+\alpha)}{4\mu\alpha}\,,\quad \varkappa^2 = \frac{\gamma+\epsilon}{4\alpha}\,.$$

Applying in Eq. $(5.2)_2$ the operator $\nabla^2 H$ and making use of relations $(5.2)_2$ and $(5.3)_1$, after elimination of the function $\underline{u}$ we obtain the equation

$$\left\{\begin{aligned} &DH\nabla^2\underline{\varphi} = \frac{1}{4\alpha\mu l^2}\,[2\alpha\nabla^2\operatorname{rot}\underline{X} - (\mu+\alpha)\nabla^2 H\underline{Y}\\ &-\frac{4\alpha^2}{\beta+2\gamma}\operatorname{grad}\operatorname{div}\underline{Y} + \frac{(\beta+\gamma-\epsilon)(\mu+\alpha)}{\beta+2\gamma}\operatorname{grad}\operatorname{div}\nabla^2\underline{Y}] \end{aligned}\right. \tag{5.5}$$

Eqs (5.4) and (5.5) are very useful in determining the functions $\underline{u}$ and $\underline{\varphi}$ due to an action of distortion in the infinite elastic space.

Let us consider some particular examples.

1. Consider the thermal distortion $\gamma_{ji}^{0} = \alpha_t \theta \, \delta_{ji}$, $\varkappa_{ji}^{0} = 0$. Then

$$X_i = -(2\mu + 3\lambda)\, \alpha_t \theta_{,i} \,, \quad Y_i = 0 \,. \tag{5.6}$$

Observe that in this case Eq. (5.4) is reduced to the equation

$$\nabla^2 u_i = m\theta_{,i} \,, \quad m = \frac{(3\lambda + 2\mu)\, \alpha_t}{\lambda + 2\mu} \,, \tag{5.7}$$

and that $\varphi_i = 0$. Introducing the potential of thermoelastic strain $u_i = \Phi_{,i}$ we transform Eq. (5.7) to the form

$$\nabla^2 \Phi = m\theta \,, \tag{5.8}$$

the solution of the latter equation is the function

$$\Phi(\underline{x}) = -\frac{m}{4\pi} \int_V \frac{\theta(\underline{\xi})\, dV(\underline{\xi})}{R(\underline{x} - \underline{\xi})} \,. \tag{5.9}$$

It is interesting to note that $\varphi_i = 0$. Thus, in the infinite space strain $\varkappa_{ji}$ and couple stress μ_{ji} do not appear. The stress σ_{ji} is now given by the formula

$$\sigma_{ji} = 2\mu(\Phi_{,ij} - \delta_{ij} \Phi_{,kk}) \,. \tag{5.10}$$

In the particular case of the thermal nucleus $\theta(\underline{x}) = \theta_0 \delta(\underline{x})$ we obtain

$$\Phi(\underline{x}) = -\frac{m\theta_0}{4\pi R(\underline{x}, 0)} \,, \quad u_i = \Phi_{,i} \,. \tag{5.11}$$

2. Assume now that $\gamma_{ji}^{0} = 0$ and $\varkappa_{ji}^{0} = \varkappa^{0}(\underline{x})\, \delta_{ji}$. Then

$$X_i = 0 \,, \quad Y_i = -(3\beta + 2\gamma)\, \varkappa_{,i}^{0} \,.$$

Eq. (5.5) yields $u_i = 0$ and Eq. (5.4) is reduced to the simpler Helmholtz equation

$$H\varphi_i = n\varkappa_{,i}^{0} \,, \quad n = \frac{3\beta + 2\gamma}{\beta + 2\gamma} \,, \quad H = \nabla^2 - \frac{1}{\nu^2} \,. \tag{5.12}$$

Introducing the potential $\varphi_i = \Omega_{,i}$ we reduce Eq. (5.12) to the form

$$H\Omega = n\varkappa^0(\underline{x}) \ . \tag{5.13}$$

The solution of the latter equation is the following:

$$\Omega(\underline{x}) = -\frac{n}{4\pi}\int_V \frac{\varkappa^0(\underline{\xi})\, e^{-R/\nu}}{R(\underline{x}-\underline{\xi})}\, dV(\underline{\xi}) \ . \tag{5.14}$$

In the considered case $u_i = 0$. The rotations φ_i are given by the formula $\varphi_i = \Omega_{,i}$. Besides the couple stress there occurs also the ordinary stress. We have

$$\mu_{ji} = 2\gamma(\Omega_{,ij} - \delta_{ij}\Omega_{,kk}), \quad \sigma_{ji} = -2\alpha\epsilon_{kji}\Omega_{,k} \ . \tag{5.15}$$

In the particular case $\varkappa^0(\underline{x}) = \delta(\underline{x})$ formula (5.14) yields

$$\Omega(\underline{x}) = -\frac{n}{4\pi}\frac{e^{-R/\nu}}{R}, \quad R = (x_1^2 + x_2^2 + x_3^2)^{1/2} \ . \tag{5.16}$$

Consider the case in which the right-hand side of Eq. (5.12) is the stress $n\varkappa^0 H(a-r)$ where $H(z)$ is the Heaviside function. Eq. (5.12)

$$H\Omega(r) = n\varkappa^0 H(a-r) \ , \quad r = (x_1^2 + x_2^2)^{1/2} \ , \quad \varkappa^0 = \text{const} \ , \tag{5.17}$$

is now axisymmetric. Consequently, in an infinite cylinder of radius a and axis $x_3 = z$ the distortion $\varkappa^0 = \text{const}$. Outside of the cylinder the distortion vanishes. Since the problem is axisymmetric, Eq. (5.17) has the form

$$\left(\frac{\partial^2}{\partial r^2} + \frac{1}{r}\frac{\partial}{\partial r} - \frac{1}{\nu^2}\right)\Omega(r) = n\varkappa^0 H(a-r) \ . \tag{5.18}$$

Applying to the above equation the Hankel integral transform, we obtain for the rotation angle φ_r the formula

$$\varphi_r = \frac{\partial\Omega}{\partial r} = A\int_0^\infty \frac{\xi J_1(\xi a)J_1(\xi r)}{\xi^2 + \frac{1}{\nu^2}}\, d\xi \ , \quad A = na\varkappa^2 \ . \tag{5.19}$$

From $(5.15)_1$ we obtain

$$\mu_{rr} = -\frac{2\gamma}{r}\frac{\partial\Omega}{\partial r}, \quad \mu_{\theta\theta} = -2\gamma\frac{\partial^2\Omega}{\partial r^2}, \quad \mu_{zz} = \beta\nabla^2\Omega \ . \tag{5.20}$$

Taking into account that

$$
\begin{cases}
P_1 = \int_0^\infty \frac{\zeta J_1(\zeta a) J_1(\zeta r)}{\zeta^2 + \frac{1}{\nu^2}} d\zeta = \begin{cases} I_1(\frac{r}{\nu}) K_1(\frac{a}{\nu}) & \text{for } 0<r<a \\ I_1(\frac{a}{\nu}) K_1(\frac{r}{\nu}) & \text{for } a<r<\infty \end{cases} \\
P_2 = \int_0^\infty \frac{\zeta^2 J_1(\zeta a) J_0(\zeta r)}{\zeta^2 + \frac{1}{\nu^2}} d\zeta = \begin{cases} \frac{1}{\nu} I_0(\frac{r}{\nu}) K_1(\frac{a}{\nu}) & \text{for } 0<r<a \\ 0 & \text{for } a<r<\infty \end{cases}
\end{cases}
\tag{5.21}
$$

we obtain

$$
\mu_{rr} = -2\gamma A \frac{P_1}{r}, \quad \mu_{\theta\theta} = -2\gamma A \left(P_2 - \frac{P_1}{r}\right), \quad \mu_{zz} = \beta A P_2 . \tag{5.22}
$$

Observe that μ_{rr} is a continuous function in the interval $0<r<\infty$, the function $\mu_{\theta\theta}$ has a discontinuity on the circle $r=a$ and μ_{zz} is different from zero in the interval $0<r<a$.

6. Compatibility equations

Consider the plane state of strain. Assume that all sources and unknown functions depend on the variables x_1, x_2. In this particular case the system of equations (1.12) is decomposed into two independent systems of compatibility equations.

The first system has the form

$$
\begin{cases}
\partial_1 \gamma'_{21} - \partial_2 \gamma'_{11} - \varkappa'_{13} = \alpha_{31} , \\
\partial_1 \gamma'_{22} - \partial_2 \gamma'_{12} - \varkappa'_{23} = \alpha_{32} , \\
\partial_1 \varkappa'_{23} - \partial_2 \varkappa'_{13} = \theta_{33} ,
\end{cases}
\tag{6.1}
$$

where

$$
\begin{cases}
\alpha_{31} = -\partial_1 \overset{0}{\gamma}_{21} + \partial_2 \overset{0}{\gamma}_{11} + \overset{0}{\varkappa}_{13} , \\
\alpha_{32} = -\partial_1 \overset{0}{\gamma}_{22} + \partial_2 \overset{0}{\gamma}_{12} + \overset{0}{\varkappa}_{23} , \\
\theta_{33} = -\partial_1 \overset{0}{\varkappa}_{23} + \partial_2 \overset{0}{\varkappa}_{13} .
\end{cases}
$$

It is readily observed that in the considered case the displacement and rotation vectors have the form

$$
\underline{u} = (u_1, u_2, 0) , \quad \underline{\varphi} = (0, 0, \varphi_3) . \tag{6.2}
$$

The system of equations (6.1) can be transformed to the form

$$\begin{cases} \partial_2^2\gamma'_{11} + \partial_1^2\gamma'_{22} - \partial_1\partial_2(\gamma'_{12} + \gamma'_{21}) = A_1 \,, \\ \partial_1\partial_2(\gamma'_{11} - \gamma'_{22}) + \partial_2^2\gamma'_{12} - \partial_1^2\gamma'_{21} + \partial_1\varkappa'_{13} + \partial_2\varkappa'_{23} = A_2 \,, \\ \partial_1\varkappa'_{23} - \partial_2\varkappa'_{13} = A_3 \,, \end{cases} \tag{6.3}$$

where

$$A_1 = \theta_{33} + \partial_1\alpha_{32} - \partial_2\alpha_{31} \,, \quad A_2 = -\partial_1\alpha_{31} - \partial_2\alpha_{32} \,, \quad A_3 = \theta_{33} \,.$$

Replacing strain γ'_{ji}, $\varkappa'_{ji}$ by stress σ'_{ji}, μ'_{ji} (making use of relations (1.2)) we arrive at a system of three equations in stresses

$$\partial_2^2\sigma_{11} + \partial_1^2\sigma_{22} - \frac{\lambda}{2(\lambda+\mu)}\nabla_1^2(\sigma_{11}+\sigma_{22}) - \partial_1\partial_2(\sigma_{12}+\sigma_{21}) = 2\mu A_1 \,,$$

$$(\partial_2^2-\partial_1^2)(\sigma_{12}+\sigma_{21}) + \frac{\mu}{\alpha}\nabla_1^2(\sigma_{12}-\sigma_{21}) + \frac{4\mu}{\gamma+\epsilon}(\partial_1\mu_{13}+\partial_2\mu_{23}) + 2\partial_1\partial_2(\sigma_{11}-\sigma_{22}) = 4\mu A_2 \,,$$

$$\partial_1\mu_{23} - \partial_2\mu_{13} = (\gamma+\epsilon)A_3 \,. \tag{6.4}$$

The state of stress has now the form

$$\underline{\sigma} = \begin{Vmatrix} \sigma_{11} & \sigma_{12} & 0 \\ \sigma_{21} & \sigma_{22} & 0 \\ 0 & 0 & \sigma_{33} \end{Vmatrix} \qquad \underline{\mu} = \begin{Vmatrix} 0 & 0 & \mu_{13} \\ 0 & 0 & \mu_{23} \\ \mu_{31} & \mu_{32} & 0 \end{Vmatrix} \tag{6.5}$$

Three of the above components, namely the stresses σ_{33}, μ_{31}, μ_{32}, can be expressed by the remaining ones. The system of equations (6.4) contain six unknown components of the state of stress.

The compatibility equations should be completed by the equilibrium equations

$$\partial_1\sigma_{11} + \partial_2\sigma_{21} = 0 \,, \quad \partial_1\sigma_{12} + \partial_2\sigma_{22} = 0 \,, \quad \partial_1\mu_{13} + \partial_2\mu_{23} + \sigma_{12} - \sigma_{21} = 0 \,. \tag{6.6}$$

Then the number of equations is equal to the number of the unknowns.

Let us consider two particular cases.

1. $A_1 \neq 0$, $A_2 \neq 0$, $A_3 = 0$.

We express the stress by the Airy-Mindlin function

$$\begin{cases} \sigma_{11}=\partial_2^2 F-\partial_1\partial_2\Psi\,, & \sigma_{22}=\partial_1^2 F+\partial_1\partial_2\Psi\,,\\ \sigma_{12}=-\partial_1\partial_2 F-\partial_2^2\Psi\,, & \sigma_{21}=-\partial_1\partial_2 F+\partial_1^2\Psi\,,\\ \mu_{13}=\partial_1\Psi\,, & \mu_{23}=\partial_2\Psi\,, \end{cases} \tag{6.7}$$

then the equilibrium equations and Eq. $(6.4)_3$ are identically satisfied and the remaining equations $(6.4)_{1,2}$ are reduced to the simple differential equations

$$\nabla_1^2\nabla_1^2 F=\frac{4\mu(\lambda+\mu)}{\lambda+2\mu}A_1\,,\quad \nabla_1^2(l^2\nabla_1^2-1)\Psi=-(\gamma+\epsilon)A_2\,. \tag{6.8}$$

Let us consider some particular cases.

a) Assume that at the origin there acts a concentrated distortion $\gamma_{11}^0(x_1,x_2)$ of intensity γ^0: $\gamma_{11}^0(x_1,x_2)=\gamma^0\delta(x_1)\,\delta(x_2)$. Then $A_1=-\gamma^0\partial_2^2\delta(x_1)\,\delta(x_2)$, $A_2=-\gamma^0\partial_1\partial_2\delta(x_1)\,\delta(x_2)$. Substituting into (6.8) and solving the equations we obtain

$$\begin{cases} F=\dfrac{\mu(\lambda+\mu)\gamma^0}{2\pi(\lambda+2\mu)}\dfrac{\partial^2}{\partial x_2^2}\,[r^2(lnr+C)]\,, & C-\text{Euler's constant}\\ \Psi=-\dfrac{(\gamma+\epsilon)\gamma^0}{2\pi}\dfrac{\partial^2}{\partial x_1\partial x_2}\,[lnr+K_0(\frac{r}{l})]\,, \end{cases} \tag{6.9}$$

where $K_0(\frac{r}{l})$ is the modified Bessel function of third kind, i.e. the Macdonald function. The stresses are calculated by means of formulae (6.7).

b) Assume that at the origin there acts the distortion $\gamma_{22}^0=\gamma^0\delta(x_1)\,\delta(x_2)$; then $A_1=\gamma^0\partial_1^2\delta(x_1)\,\delta(x_2)$, $A_2=\gamma^0\partial_1\partial_2\delta(x_1)\,\delta(x_2)$.

Solving Eqs (6.8) we obtain the following particular integrals:.

$$\begin{cases} F=\dfrac{\mu(\lambda+\mu)\gamma^0}{2\pi(\lambda+2\mu)}\dfrac{\partial^2}{\partial x_1^2}\,[r^2(lnr+C)]\,,\\ \Psi=\dfrac{(\gamma+\epsilon)\gamma^0}{2\pi}\dfrac{\partial^2}{\partial x_1\partial x_2}\,(lnr+K_0(\frac{r}{l}))\,. \end{cases} \tag{6.10}$$

Observe that when $\gamma_{ji}^0=\gamma^0\delta(x_1)\,\delta(x_2)\,\delta_{ji}$, adding (6.9) and (6.10) we have

$$F=\frac{\mu(\lambda+\mu)\gamma^0}{2\pi(\lambda+2\mu)}\nabla_1^2[r^2(lnr+C)]=\frac{2\mu(\lambda+\mu)}{\pi(\lambda+2\mu)}(lnr+C)\,,\quad \Psi=0\,. \tag{6.11}$$

This case can be interpreted as the action of a temperature nucleus at the origin. Observe that in this case the couple stresses $\mu_{13}, \mu_{23}, \mu_{31}, \mu_{32}$ vanish.

2. Consider now the case of the distortion $\varkappa^0_{13} = -\varkappa^0_{23} = \varkappa^0 \delta(x_1)\,\delta(x_2)$. Thus, we have

$$A_1 = 0\,,\; A_2 = 0 \quad A_3 = (\partial_1 + \partial_2)\,\varkappa^0 \delta(x_1)\,\delta(x_2)\ .$$

Now the representation of stress in terms of the Airy-Mindlin function is unsuitable. Eliminating from Eqs (6.5) and (6.6) in turn the stresses and taking into account the equilibrium equations, we arrive at the system of equations

$$\begin{aligned}
&\nabla_1^2 \mu_{13} = -(\gamma+\epsilon)\,\partial_2 A_3\,, \quad \nabla_1^2 \mu_{23} = (\gamma+\epsilon)\,\partial_1 A_3\,,\\
&\nabla_1^2 \nabla_1^2 \sigma_{11} = p\,\partial_2^2 A_3\,, \quad \nabla_1^2 \nabla_1^2 \sigma_{22} = p\,\partial_1^2 A_3\,,\\
&\nabla_1^2 \nabla_1^2 \sigma_{12} = -p\,\partial_1 \partial_2 A_3\,, \quad \sigma_{21} = \sigma_{12}\,, \quad p = \frac{4\mu(\lambda+\mu)}{\lambda+2\mu}\ .
\end{aligned} \tag{6.12}$$

Solving them we have

$$\begin{aligned}
&\mu_{13} = \frac{1}{2\pi}(\gamma+\epsilon)\,\varkappa^0 \partial_2(\partial_1+\partial_2)\,I_1\,, \quad \mu_{23} = -\frac{1}{2\pi}(\gamma+\epsilon)\,\varkappa^0 \partial_1(\partial_1+\partial_2)\,I_1\,,\\
&\sigma_{11} = \frac{p\varkappa^0}{2\pi}\,\partial_2^2(\partial_1+\partial_2)\,I_2\,, \quad \sigma_{22} = \frac{p\varkappa^0}{2\pi}\,\partial_1^2(\partial_1+\partial_2)\,I_2\,,\\
&\sigma_{12} = -\frac{p\varkappa^0}{2\pi}\,\partial_1\partial_2(\partial_1+\partial_2)\,I_2\,, \quad \sigma_{21} = \sigma_{12}\,,
\end{aligned} \tag{6.13}$$

where

$$I_1 = -(\ln r + C)\,, \quad I_2 = r^2(\ln r + C)\ .$$

Let us now examine the second system of compatibility equations related to the so-called second problem of plane state of strain, when

$$\underline{u} = (0, 0, u_3)\,, \quad \underline{\varphi} = (\varphi_1, \varphi_2, 0)\ . \tag{6.14}$$

This system has the form

$$\begin{cases} \partial_2 \gamma'_{31} + \varkappa'_{22} = \alpha_{11}, & -\partial_1 \gamma'_{32} + \varkappa'_{11} = \alpha_{22}, \\ \partial_2 \gamma'_{32} - \varkappa'_{21} = \alpha_{12}, & -\partial_1 \gamma'_{31} - \varkappa'_{12} = \alpha_{21}, \\ \partial_1 \gamma'_{23} - \partial_2 \gamma'_{13} + \varkappa'_{11} + \varkappa'_{22} = \alpha_{33}, \end{cases} \tag{6.15}$$

where

$$\begin{cases} \alpha_{11} = -\partial_2 \overset{0}{\gamma}_{31} - \overset{0}{\varkappa}_{22}, & \alpha_{22} = \partial_1 \overset{0}{\gamma}_{32} - \overset{0}{\varkappa}_{11}, \\ \alpha_{12} = -\partial_2 \overset{0}{\gamma}_{32} + \overset{0}{\varkappa}_{21}, & \alpha_{21} = \partial_1 \overset{0}{\gamma}_{31} + \overset{0}{\varkappa}_{12}, \\ \alpha_{33} = -\partial_1 \overset{0}{\gamma}_{23} + \partial_2 \overset{0}{\gamma}_{13} - \overset{0}{\varkappa}_{11} - \overset{0}{\varkappa}_{22} . \end{cases}$$

Making use of the constitutive relations (1.3) we transform Eqs (6.15) to the form in which only stresses occur:

$$\begin{cases} \partial_1^2 \mu_{22} + \partial_2^2 \mu_{11} - \dfrac{\beta}{2(\gamma+\beta)} \nabla_1^2 (\mu_{11} + \mu_{22}) - \partial_1 \partial_2 (\mu_{12} + \mu_{21}) = 2\gamma B_1 , \\ (\partial_2^2 - \partial_1^2)(\mu_{12} + \mu_{21}) + \dfrac{\gamma}{\epsilon} \nabla_1^2 (\mu_{12} - \mu_{21}) - 2\partial_1 \partial_2 (\mu_{22} - \mu_{11}) = 4\gamma B_2 , \\ \partial_1 (\sigma_{23} + \sigma_{32}) - \partial_2 (\sigma_{31} + \sigma_{13}) = 2\mu B_3 , \\ \mu_{11} + \mu_{22} + \dfrac{\gamma+\beta}{\alpha} (\partial_2 \sigma_{31} - \partial_1 \sigma_{32}) = (\gamma+\beta)\left(2B_5 - \dfrac{\mu+\alpha}{\alpha} B_3\right) , \\ \mu_{12} - \mu_{21} - \dfrac{\epsilon(\mu+\alpha)}{2\mu\alpha} (\partial_1 \sigma_{31} + \partial_2 \sigma_{32}) = 2\epsilon B_4 , \end{cases} \tag{6.16}$$

where

$$B_1 = \partial_1^2 \alpha_{11} + \partial_2^2 \alpha_{22} + \partial_1 \partial_2 (\alpha_{11} + \alpha_{21}) ,$$

$$B_2 = \partial_1 \partial_2 (\alpha_{22} - \alpha_{11}) + \partial_2^2 \alpha_{21} - \partial_1^2 \alpha_{12} ,$$

$$B_3 = \alpha_{33} - (\alpha_{11} + \alpha_{22}) , \quad B_4 = \alpha_{21} - \alpha_{12}, \quad B_5 = \alpha_{33} .$$

The system of five equations (6.16) contains eight unknown stresses. Completing the compatibility equations (6.16) by the three equilibrium equations

$$\partial_1 \mu_{11} + \partial_2 \mu_{12} + \sigma_{23} - \sigma_{32} = 0 , \quad \partial_1 \mu_{12} + \partial_2 \mu_{22} + \sigma_{31} - \sigma_{13} = 0 ,$$

$$\partial_1 \sigma_{13} + \partial_2 \sigma_{23} = 0 , \tag{6.17}$$

we obtain a system of equations, the number of which is equal to the number of the unknowns.

In the particular case of the distortion $\gamma^0_{23} = \gamma^0 \delta(x_1)\,\delta(x_2)$, the remaining distortions being equal to zero, we have

$$B_1 = B_2 = B_4 = 0\ , \quad B_3 = B_5 = -\gamma^0 \partial_1 \delta(x_1)\,\delta(x_2)\ .$$

Eliminating in turn the stresses from Eqs (6.16) and (6.17) we obtain the following equations for the stresses:

$$\begin{cases} D\nabla_1^2 \sigma_{13} = -\partial_2 N A_5\ , \quad D\nabla_1^2 \sigma_{23} = \partial_1 N A_5\ , \\ D\nabla_1^2 \sigma_{31} = -\partial_2 M A_5\ , \quad D\nabla_1^2 \sigma_{32} = \partial_1 M A_5\ . \end{cases} \tag{6.18}$$

Here

$$N = (\alpha+\mu)\,D+\alpha\ , \quad M = (\mu-\alpha)\,D-\alpha\ , \quad D = l^2\nabla_1^2 - 1\ , \quad A_5 = -\gamma^0 \partial_1 \delta(x_1)\,\delta(x_2)\ .$$

Solving the above equations we obtain

$$\begin{cases} \sigma_{13} = \dfrac{\gamma^0}{2\pi}\,\partial_1\partial_2(\mu I_1 + \alpha \overset{\nu}{I}_2)\ , \qquad \sigma_{31} = \dfrac{\gamma^0}{2\pi}\,\partial_1\partial_2(\mu I_1 - \alpha \overset{\nu}{I}_2)\ , \\ \sigma_{23} = -\dfrac{\gamma^0}{2\pi}\,\partial_1^2(\mu I_1 - \alpha \overset{\nu}{I}_2)\ , \qquad \sigma_{32} = -\dfrac{\gamma^0}{2\pi}\,\partial_1^2(\mu I_1 + \alpha \overset{\nu}{I}_2)\ , \end{cases} \tag{6.19}$$

where

$$I_1 = -(\ln r + C)\ , \quad \overset{\nu}{I}_2 = K_0\left(\frac{r}{\nu}\right)\ , \quad \nu = \left(\frac{2\gamma+\beta}{4\alpha}\right)^{1/2}\ .$$

Eliminating the stresses from Eqs (6.16) and (6.17) and making use of Eqs (6.18) we arrive at the following system of differential equations for the couple stresses:

$$\begin{aligned} &D\nabla_1^2 \mu_{22} = -\left(\gamma\partial_2^2 + \frac{\beta}{2}\nabla_1^2\right) A_5\ , \\ &D\nabla_1^2 \mu_{11} = -\left(\gamma\partial_1^2 + \frac{\beta}{2}\nabla_1^2\right) A_5\ , \\ &D\nabla_1^2 \mu_{12} = -2\gamma\partial_1\partial_2 A_5\ , \quad \mu_{21} = \mu_{12}\ , \quad A_5 = -\gamma^0 \partial_1 \delta(x_1)\,\delta(x_2)\ . \end{aligned} \tag{6.20}$$

Solving them we obtain the formulae

$$\left\{\begin{aligned} \mu_{11} &= -\frac{\overset{0}{\gamma}}{2\pi}\,\partial_1\left[\frac{\beta}{\nu^2}\overset{\nu}{I}_2 - 2\gamma\partial_1^2(I_1 - \overset{\nu}{I}_2)\right] , \\ \mu_{22} &= -\frac{\overset{0}{\gamma}}{2\pi}\,\partial_1\left[\frac{\beta}{\nu^2}\overset{\nu}{I}_2 - 2\gamma\partial_2^2(I_1 - \overset{\nu}{I}_2)\right] , \\ \mu_{12} = \mu_{21} &= -\frac{\gamma\overset{0}{\gamma}}{\pi}\,\partial_1^2\partial_2(I_1 - \overset{\nu}{I}_2) . \end{aligned}\right. \tag{6.21}$$

A determination of the singular solutions (the Green functions) for the distortion components makes it possible to calculate by integration the state of stress due to distortions distributed over an arbitrary bounded region Γ .

References

1. NOWACKI, W.: Theory of micropolar elasticity, CISM Udine; Wien: Springer. 1970.
2. ANTHONY, K.H.: Die Theorie der Disklinationen. Arch. for Rat. Mech. Analysis, 39, 43–88 (1970).
3. CLAUS, W.D., and ERINGEN, A.C.: Tree dislocation concepts and micromorphic mechanics. Development in Mechanics, 6, Proc. 12th Midwestern Mechanics Conference 1969.
4. MAYSEL, V.M.: Temperature problem of the theory of elasticity (in Russian). Kiev: 1951.
5. SANDRU, N.: On some problems of the linear theory of the asymmetric elasticity. Int. J. Engng. Sci. 4, 81–94 (1966).

THE YIELD CRITERION IN THE GENERAL CASE OF NONHOMOGENEOUS STRESS AND DEFORMATION FIELDS

J.A. KÖNIG, Warsaw and W. OLSZAK, Warsaw-Udine

1. Introduction

The transition of a material into the plastic state is connected with a change of its structure. It is obvious that, even within the framework of a continuum theory, one has to take into account the fact that the notion of structure does make sense only if it is related to a finite region (and not to a single point), and also, in an analogous way, to a finite interval of time.

Thus, if the behaviour of the considered body at a given instant and in a given point has to be described, one has to assume that this behavior will explicitely depend also on prior states (in time) as well as on states of neighboring points (in space).

As the notions of spatial and time neighborhood are not precise, it seems to be justified to introduce, as a first approximation, the dependence on the corresponding gradients; hence, in our case, the dependence on the gradients of the stress and deformation tensors with respect to their space and time coordinates, as these gradients characterize the changes of state in the spatial neighborhood of the point as well as in neighboring instants.

That such influences actually do exist has been proved by experimental evidence. The corresponding observed facts are:

(a) the phenomenon of the so called "overelasticity" (see

e.g. /4/, /7/, /8/)[1] resulting in an increase of the yield limit if the state of stress is spatially not homogeneous;

(b) the phenomenon of a pronounced dependence of the yield limit on the rate of loading and deformation (e.g. /1o/), usually explained by the viscous material properties.

The today's intense research activities in the field of viscoplasticity proceed from the facts mentioned under (b), whereas it seems that the influence of spatial nonhomogeneity of stress and deformation fields on the formulation of the yield condition has not been considered up to now.

The first engineering attempts to assess the phenomena mentioned under (a) date back to the years before World War II. The relevant results are incomplete and difficult to obtain. Instructive coherent experimental research programmes have been conceived and implemented, as far as we know, only in the last decade by F. Campus and his Belgian coworkers. They will be referred to later on.

Incidentally, it may be mentioned that phenomenologically apparently homogeneous states of stress cannot be considered to be strictly such in reality, in consideration of the granular structure of materials. The well known "region of instability" in the classical stress-strain diagram when reaching the yield limit can be attributed to similar reasons, although uniaxial tension usually is considered as a particularly characteristic example of a homogeneous state of stress. These facts have been known for a long time; Fig. 1, taken from L. Prandtl's paper /12/, allows for this phenomenon in an idealized manner.

All these experimentally observed results can be theoretically represented with relative ease if the yield condition is assumed to depend also on the <u>changes</u> of stress and deformation fields. We shall consider these changes to be quite general, i.e., as changes with respect to spatial coordinates as well as with respect to time.

---

[1] In French "surélasticité". Purposely we avoid other expressions (like, e.g., combinations with "hyper..." or "super...", etc.), which already are in current use to designate other phenomena.

For simplicity of presentation, a four-dimensional time-space system is introduced, but this being to be understood in the sense of classical Mechanics. Consequently, the (absolute) time t becomes a privileged coordinate. Thus as the possible transformations of the reference system we shall consider merely the following ones:

$$x'_\alpha = x_\alpha(x_1, x_2, x_3), \qquad (\alpha = 1,2,3)$$
$$x'_4 = x_4+t_o. \qquad (1)$$

Within the framework of such tranformations, the classical quantities σ_{ij}, ε_{ij} (the stress and strain tensors, for small deformations, respectively) will preserve their tensorial properties. They will transform under a change of the reference system like tensors of rank two:

$$\sigma'_{ij} = A_{ki}A_{lj}\sigma_{kl}; \quad \varepsilon'_{ij} = A_{ki}A_{lj}\varepsilon_{kl}; \quad (k,l = 1,2,3,4)$$
$$A_{kl} = \frac{\partial x'_k}{\partial x_i}, \qquad (2)$$

if $\sigma_{4i} = \sigma_{i4} = \varepsilon_{4i} = \varepsilon_{i4} = 0$ is assumed. Their gradients $\sigma_{ij,k}$, $\varepsilon_{ij,k}$ are tensors of rank three.

Accordingly, we shall assume that the transition of the material into the plastic state depends not only on the tensors σ_{ij}, ε_{ij} themselves, as usually supposed, but also on the gradients mentioned above. This dependence can be of various types and we will consider especially the following ones:

(1) an explicite dependence on the components of the gradients;

(2) a dependence on the invariants of the system of tensors σ_{ij}, ε_{ij}, $\sigma_{ij,k}$, $\varepsilon_{ij,k}$;

(3) a dependence on the gradients of some quantities which are scalar functions of the tensors σ_{ij}, ε_{ij};

(4) a dependence on the derivatives of these scalar functions with respect to the components of the tensors σ_{ij}, ε_{ij}.

It can be shown quite easily that, for isotropic bodies, (1) and (2) are equivalent. Furthermore, (3) is a special case of (2), inasmuch as a scalar function of the gradient of a scalar function of the tensors σ_{ij}, ε_{ij} is obviously a scalar function of the invariants of the system of tensors σ_{ij}, ε_{ij}, $\sigma_{ij,k}$, $\varepsilon_{ij,k}$; as a matter of fact, we have

$$f\left[g(\sigma_{ij},\varepsilon_{ij})_{,k}\right]=f\left[\frac{\partial g}{\partial\sigma_{ij}}\sigma_{ij,k}+\frac{\partial g}{\partial\varepsilon_{ij}}\varepsilon_{ij,k}\right]=f_1\left[\sigma_{ij},\varepsilon_{ij},\sigma_{ij,k},\varepsilon_{ij,k}\right];$$

furthermore f_1, being a scalar function of the system of tensors σ_{ij}, ε_{ij}, $\sigma_{ij,k}$, $\varepsilon_{ij,k}$, must be a function of the invariants of this system of tensors only.

Finally, case (4) leads back to the classical theory, i.e. to the yield condition of the from

$$\Psi(\sigma_{ij},\ \varepsilon_{ij}) = 0.$$

as can be shown by a detailed analysis.

2. The Plasticity Condition

Following the pattern of these considerations, the general form of the yield condition will be assumed to have the form

$$\Psi(\sigma_{ij},\ \varepsilon_{ij},\ \alpha_{ijk},\ \beta_{ijk}) = 0 \qquad (i,j,k = 1,2,3,4) \qquad (3)$$

where $\alpha_{ijk} = \sigma_{ij,k}$, $\beta_{ijk} = \varepsilon_{ij,k}$ are tensors of rank three, each having 64 components.

All these components cannot be independent if condition (2) as well as other relcvant physical relations are taken into account; as a matter of fact:

(1) the tensors σ_{ij}, ε_{ij} are symmetric; thus $\alpha_{ijk} = \alpha_{jik}$, $\beta_{ijk} = \beta_{jik}$ (which gives 24 relations in the components);

(2) the tensor σ_{ij} has to satisfy the equations of motion

$$\sigma_{ij,j} + F_i = \rho\ddot{u}_i,$$

and the tensor ε_{ij} the compatibility **relations**

$$\varepsilon_{ij,kl} + \varepsilon_{kl,ij} - \varepsilon_{il,jk} - \varepsilon_{jk,il} = 0;$$

(this gives 3 additional relations in each of the above cases);

(3) because of $\sigma_{4i} = \varepsilon_{4i} = 0$, one has $\alpha_{i4k} = \beta_{i4k} = 0$ (which leads in addition to 16 equations).

Therefore, each of the tensors α_{ijk}, β_{ijk} has 18 independent components. Since in each system of 18 independent components their values can be calculated in an arbitrary reference system which satisfies relations (1), the complete system of independent invariants of each of these tensors must contain 18 quantities. Thus, for an isotropic material, the function Ψ of Eq. (3) will in general depend on 42 variables.

3. Special Cases of the Yield Condition

In view of the complicated form of relation (3) and especially with respect to real applications, it seems to be appropriate to introduce and analyze expressions of simpler form. This will be all the more justified as the influence of the various variables, occurring in the function Ψ, does not seem to be of equal importance, and thus for practical purposes a good approximation of Ψ may contain much less of them.

The first approximation will refer to the notion of ideal plasticity; then the function Ψ reduces to

$$\Psi(\sigma_{ij}, \alpha_{ijk}) = 0.$$

Furthermore, in agreement with the considerations of the preceeding Section, which limits the number of the considered invariants, we will investigate the following form of the function Ψ:

$$\Psi(I_1, I_2, I_3, g_1, \ldots, g_n) = 0; \qquad (4)$$

here I_1, I_2, I_3 denote the invariants of the stress tensor, whereas in the invariants $g_1, \ldots, g_n$ also the components of the gradient tensor shall occur.

The special form of transformation (1) permits to consider the invariant condition (4) for $n = 1$, where $g_1 = g$ shall depend solely on $I_i (i = 1,2,3)$ and their derivatives with respect to $x_i (i = 1,2,3,4)$.

This relatively simple form allows already to describe phenomena reflecting "overelasticity" (a) as well as those due to "rate sensitivity" (b). An obvious and simple assumption would be to put

$$g = f_{,i} \cdot f_{,i} \qquad (i = 1,2,3,4)$$

where f is a scalar function of the stress tensor. This gives finally

$$\Psi = (I_1, I_2, I_3, g) = 0. \qquad (5)$$

This expression takes formally the form of a yield condition which explicitely depends on a certain parameter (like temperature or time) (See, e.g., /1/, /11/). In Sec. 5, where loading and unloading conditions are discussed, we will make use of this analogy.

To assess in detail the possibilities, which are represented by condition (5), let us consider one of its particular cases, which leads to a generalization of the yield condition usually used for metals, namely the form

$$\Psi = \varphi(I_2, I_3) - c(g).$$

For the Huber-v. Mises condition there follows

$$\Psi = \frac{1}{2} s_{ij}s_{ij} - c(g); \qquad (6)$$

$$g = s_{ij}s_{ij,k}s_{mn}s_{mn,k}, \qquad (i,j,m,n, = 1,2,3,4),$$

where s_{ij} denotes the stress deviator.

It seems to be appropriate to assume the function c(g) to equate k^2 (k being the yield limit in simple shear) for g = 0, and to remain bounded for arbitrarily great gradients. Such properties has, e.g., the function

$$c(g) = k^2 + (K^2 - k^2)\vartheta(g), \qquad (7)$$

where

$$K > k, \quad \vartheta(0) = 0, \quad \vartheta(\infty) = 1.$$

4. Example: Pure Bending

As a simple example, which illustrates our considerations, a slowly developing process of pure bending of a rectangular beam is treated next. This example corresponds to most of the experimental arrangements mentioned in Sec. 1 (see Fig. 2). The material of the beam shall obey condition (6) with c(g) being given by expression (7).

It is obvious that $\sigma_y = \sigma(x)$ is the only nonvanishing stress component with

$$\sigma(x) = \sigma(H)\,\frac{x}{H} = \frac{3}{2}\,\frac{M}{BH^3}\,x; \qquad \sigma(H) = \frac{3}{2}\,\frac{M}{BH^2},$$

in the purely elastic state. The only nonvanishing component of the stress gradient is

$$\sigma(x)_{,x} = \frac{\sigma(H)}{H} = \frac{3}{2}\frac{M}{BH^3} .$$

Consequently, the condition (6) can be expressed in the following form:

$$\frac{2}{9}\sigma^2 - k^2 - (K^2-k^2)\vartheta(\frac{\sigma(H)}{H}) = 0$$

or

$$\Psi = \frac{1}{2}(\frac{M}{BH^3})^2 x^2 - k^2 - (K^2-k^2)\vartheta(\frac{3M}{2BH^3}) = 0.$$

One can see that Ψ takes on its maximum value for $x = \pm H$. Thus, the first plastic deformations will occur in the outer layers for $M \doteq M^*$, where M^* can be found as the root of the equation

$$\frac{1}{2}(\frac{M}{BH^3})^2 H^2 - k^2 - (K^2-k^2)\vartheta(\frac{3M}{2BH^3}) = 0.$$

A further load increase leads to the development of plastic regions, delimited by $\eta \le x \le H$ and $-H \le x \le \eta$.

It is evident that $M^* > M_e$, where M_e denotes the greatest elastic bending moment as given by the classical theory. This follows from $K > k$ and the mentioned properties of the function ϑ.

Taking into account the results of his interesting and carefully conducted experiments, F. Campus advances the opinion /4/ that $M^* = M_o$, i.e.

$$M^* = \sigma_H \frac{2}{3} BH^2 = \sigma_o BH^2, \tag{8}$$

where M_o denotes the value of the ultimate (limit) moment; at the instant of reaching this moment a sudden redistribution of stresses in the cross section occurs which, according to /4/, leads to the known limit stress distribution corresponding to the classical plasticity theory (see Fig. 3).

Such a process would however be in contradiction to the theory developed above. Because of Fig. 3 we would have in the point $x = H$

$$\sigma = \sigma_H, \quad \sigma_{,x} = \frac{\sigma_H}{H};$$

because of Eq. (8), one should have $\sigma_H = \frac{3}{2}\sigma_o$ which, by insertion into Eq. (5), would yield

$$\Psi(\frac{3}{2}\sigma_o, 0, 0, \frac{3\sigma_o}{2H}) = 0.$$

But this is in contradiction to the assumption of the dependence of the function on the stress gradient. Thus, it seems that, in presence of "overelasticity", i.e. when the stress gradient induces an increase of the yield limit, the considered process must develop in a somewhat more complicated manner (Fig. 4).

The stress gradient does indeed lead to an increase of the yield limit. This means that, in fact, we obtain elastically greater bending moments than would correspond to the classical elastic limit moment (and this in accordance with F. Campus' experiments). If $M = M^*$ is reached, the outer zones become plastic. To stay close to the existing conceptions, it seems to be appropriate to assume σ = const. in the plastic region, and thus one should have $\sigma = \sigma_o$. This then would result in the sudden redistribution of stresses as shown in Figs. 4c, d.

The further process under ulteriorly increasing loading would already follow the classical plasticity theory. The difference to the latter would only be that the ratio of the greatest elastically reached bending moment to the ultimate one is not 2/3, but depends on the height of the beam. For very high beams it approaches 2/3, but for low ones 1. The latter is in agreement with F. Campus' experimental results which, for such cases, showed an "almost complete" absence of regions of permanent plastic strains. From the data given in /4/, one can conclude that the used test samples have been rather thin. This would explain the mentioned almost complete absence of irreversible deformations after unloading, as well as the concentration of plastic strains to very narrow regions.

Admittedly, also this explanation is not quite satisfactory and needs further investigation. This will be done in a further study, containing a critical discussion of the problems from the theoretical point of view and a confrontation with available experimental data.

5. Criteria for Neutral, Passive and Active Processes

If the yield criterion is assumed to be of the form (3), this means that states with $\Psi > 0$ are impossible. Thus we have

$$\dot{\Psi} \leq 0, \qquad \text{if} \quad \Psi = 0. \tag{9}$$

Let us consider an element, the state of which is represented in the stress space by a point always remaining on the yield surface (3). The rate of change of Ψ is given by

$$\Psi_{,4} = \dot{\Psi} = \frac{\partial \Psi}{\partial \sigma_{ij}} \dot{\sigma}_{ij} + \frac{\partial \Psi}{\partial \varepsilon_{ij}} \dot{\varepsilon}_{ij} + \frac{\partial \Psi}{\partial \alpha_{ijk}} \dot{\alpha}_{ijk} + \frac{\partial \Psi}{\partial \beta_{ijk}} \dot{\beta}_{ijk} \tag{10}$$

$$(i,j,k = 1,2,3,4)$$

Processes with plastic deformations (for which $\dot{\varepsilon}_{ij}{}^{p} \neq 0$), can take place only if $\Psi = 0$ as well as $\dot{\Psi} = 0$. But the inverse conclusion is, in general, not valid. Processes for which

$$\dot{\varepsilon}_{ij}{}^{p} = 0, \qquad \Psi = 0, \qquad \dot{\Psi} = 0 \tag{11}$$

are called <u>n e u t r a l</u> changes (processes). Thus we obtain in an explicit form

$$\frac{\partial \Psi}{\partial \sigma_{ij}} \dot{\sigma}_{ij} + \frac{\partial \Psi}{\partial \alpha_{ijk}} \dot{\alpha}_{ijk} + \frac{\partial \Psi}{\partial \beta_{ijk}} \dot{\beta}_{ijk} = 0. \tag{12}$$

If in a certain region a neutral process takes place, then, naturally, one has

$$\dot{\varepsilon}_{ij,k} = \dot{\beta}_{ijk} = \beta_{ijk,4} = 0$$

there.

If, because of a change of the state of stress, the value of the function decreases, this change will take place purely elastically. For such a process, called an <u>u n l o a d i n g</u> (or <u>p a s s i v e</u>) process, we have

$$\frac{\partial \Psi}{\partial \sigma_{ij}} \dot{\sigma}_{ij} + \frac{\partial \Psi}{\partial \alpha_{ijk}} \dot{\alpha}_{ijk} + \frac{\partial \Psi}{\partial \beta_{ijk}} \dot{\beta}_{ijk} < 0. \tag{13}$$

Finally, the most interesting case in theorical investigations of plasticity can occur which corresponds to the <u>l o a d i n g</u> (or <u>a c t i v e</u>) process and for which

$$\frac{\partial \Psi}{\partial \sigma_{ij}} \dot{\sigma}_{ij} + \frac{\partial \Psi}{\partial \alpha_{ijk}} \dot{\alpha}_{ijk} + \frac{\partial \Psi}{\partial \beta_{ijk}} \dot{\beta}_{ijk} > 0. \tag{14}$$

It may be mentioned that in the expressions (1o) through (14), in $\dot{\alpha}_{ijk}$, $\dot{\beta}_{ijk}$ the second time derivatives of the componets of stress and strain tensors occur. As a matter of fact, for k = 4 we find

$$\dot{\alpha}_{ijk} = \alpha_{ij4,4} = \ddot{\sigma}_{ij},$$
$$\dot{\beta}_{ijk} = \beta_{ij4,4} = \ddot{\beta}_{ij}.$$

Consequently, the considered criteria of neutral as well as of passive and active processes depend, besides on velocities, also on accelerations.

This indicates that, when formulating the local conditions, which are related to the history of loading and which guarantee the activity of the process, these involve not only the first but also the second time derivatives.

Therefore, also the investigation of uniqueness problems, which are related to the classes of materials considered in this study, will essentially differ from the classical scheme.

6. The Flow Law

In theoretical investigations of plasticity it is usually assumed that the yield surface represents also the yield potential. Thus the plastic strain rate can be expressed in the usual manner by

$$\dot{\varepsilon}_{ij}{}^{p} = \Lambda \frac{\partial f}{\partial \sigma_{ij}}, \quad \Lambda \geq 0, \tag{15}$$

where $f(\sigma_{ij}, \varepsilon_{ij}{}^{p}, t, \theta, \ldots)$ represents the condition of the transition of the material into the plastic state (for unloading processes we have $\Lambda = 0$).

In the case of ideally plastic materials, the factor Λ is undetermined, whereas for hardening phenomena it is obtained from the condition that, for an active process ($\dot{\varepsilon}_{ij}{}^{p} \neq 0$), the point which characterizes the state has to remain on the yield surface. Thus, introduction of (15) into (1o) leads to

$$\dot{\Psi} = 0 = \frac{\partial \Psi}{\partial \sigma_{ij}} \dot{\sigma}_{ij} + \frac{\partial \Psi}{\partial \varepsilon_{ij}} \dot{\varepsilon}_{ij} + \frac{\partial \Psi}{\partial \alpha_{ijk}} \dot{\alpha}_{ijk} + \frac{\partial \Psi}{\partial \beta_{ijk}} \dot{\beta}_{ijk} =$$
$$= \frac{\partial \Psi}{\partial \sigma_{ij}} \left(\dot{\sigma}_{ij} + \Lambda \frac{\partial \Psi}{\partial \varepsilon_{ij}} + \Lambda_{,k} \frac{\partial \Psi}{\partial \beta_{ijk}} \right) + \frac{\partial \Psi}{\partial \alpha_{ijk}} (\dot{\sigma}_{ij})_{,k} + \Lambda \frac{\partial \Psi}{\partial \beta_{ijk}} \left(\frac{\partial \Psi}{\partial \sigma_{ij}} \right)_{,k} \tag{16}$$

If stresses σ_{ij}, plastic strains ε_{ij}^{p} as well as their gradients α_{ijk} and β_{ijk} are known, and if, furthermore, the rates of changes of stresses $\dot{\sigma}_{ij}$ and their gradients $\dot{\alpha}_{ijk}$ are given (because of α_{ij4}, knowledge of the acceleration of the stress tensor is required), the factor Λ can be found by solving the differential equation (16) with respect to Λ (when taking $\Psi \equiv 0$ into account).

If, on the other hand, the yield condition is independent of the deformation gradient β_{ijk}, Eq. (16) degenerates into an algebraic relation, from which the factor Λ can be immediately found in the form

$$\Lambda = -\frac{\dfrac{\partial \Psi}{\partial \alpha_{ijk}} (\dot{\sigma}_{ij})_{,k} + \dfrac{\partial \Psi}{\partial \sigma_{ij}} \dot{\sigma}_{ij}}{\dfrac{\partial \Psi}{\partial \varepsilon_{ij}^{p}} \cdot \dfrac{\partial \Psi}{\partial \sigma_{ij}}}. \tag{17}$$

R e f e r e n c e s

1. Boley, B.A. and Weiner, J.E.: Theory of Thermal Stresses, New York: Wiley, 196o .
2. Bonder, J.: Sur l'invariance du system différentiel de la dynamique des gaz réels dans l'espace-temps affine, sans métrique, AMS, 16, N.2, (1964).
3. Campbell, J.D. and Harding, J.: The effect of grain size, rate of strain and neutron irradiation on the tensile strength of iron. Response of Metals to High Velocity Deformation, p. 51, New York: Interscience, 1966.
4. Campus, F.: Plastification de l'acier doux en flexion plane simple. Bull. de la Classe des Sciences, Série 5, 49, 4, (1963).
5. Campus, F.: Plastification de l'acier doux en flexion composée. Bull. de la Classe des Sciences, Série 5, 49, 4, (1963).
6. Campus, F.: La plastification de l'acier doux en flexion simple et composée et ses effets sur flambage par compression des pièces droites élastoplastiques. Bull. de la Classe des Sciences, Série 5, 49, 5, (1963).
7. Campus, F. and Gamski, K.: Abaissement de la limite apparente d'élasticité des aciers par fluage après une amorce d'écrouissage à température ordinaire. C.R.Ac.Sci. de Paris (1955).
8. Dehousse, N.M.: Note relative à un phénomène de surélasticité en flexion constaté lors d'un essai de barreau en acier doux. Bull. de la Classe des Sciences, Série 5, 48; (1962).
9. Föppl, L.: Eine neue elastische Materialkonstante, Ing. Archiv, 7, Heft 4 (1936).
1o. Maiden, C. and Campbell, J.D.: The static and dynamic strength of a carbon steel at low temeratures, Phil. Mag., 3, 872, (1958).
11. Olszak W. and Perzyna, P.: The constitutive equations of the flow theory for a non-stationary yield condition. Proc. 11th Int. Congr. Applied Mechanics, Munich 1964, Berlin: Springer. 1966.
12. Prandtl, L.: Spannungsverteilung in plastischen Körpern.Proc. Intern. Congress Appl.Mech. Delft 1924, 43-54, (1925).

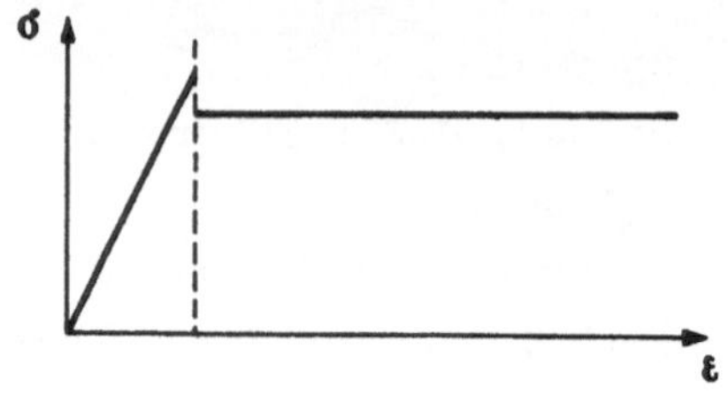

Fig. 1

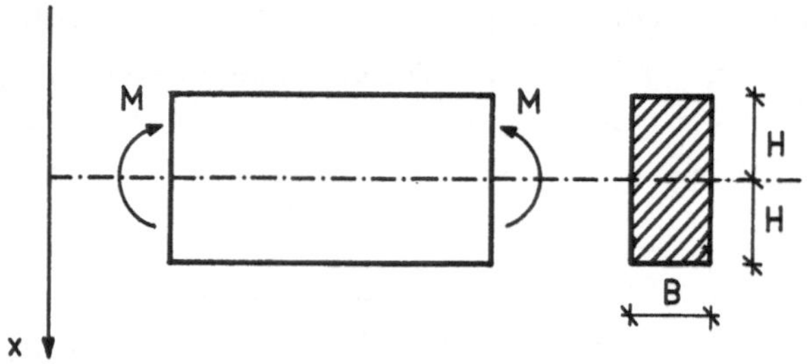

Fig. 2

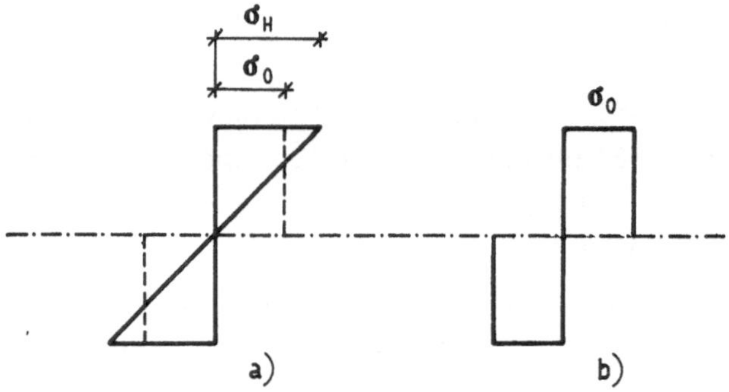

Fig. 3

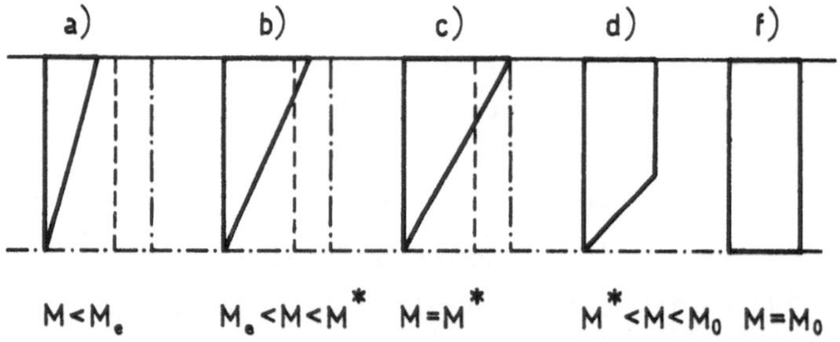

Fig. 4

ELECTRO-MAGNETO-ELASTICITY

J.B. ALBLAS, Eindhoven

1. Introduction

In this paper we consider the finite deformation of a magnetized and/or polarized elastic body, placed in an electro-magnetic field. We shall be concerned with the general dynamic problem, in which the body may be magnetically saturated or non-saturated, and may conduct currents. But we shall confine our considerations to the non-relativistic case, i.e. we assume the velocities to be small with respect to the velocity of light. In our theory we aim at a unification of electro-magnetic theory and continuum mechanics: we shall derive the force distribution of electro-magnetic origin and the influence of the deformation of the body on the field.

The literature on the interactions of electric and/or magnetic fields with elastic media is extensive. Toupin, in a pioneering contribution [1] has given an investigation of the interaction of an electrostatic field with a perfectly elastic dielectric. In a subsequent paper [2] he extended his results to include the effects of motion. For the magnetic case Tiersten [3], [4] has developed a similar theory for the deformation of magnetically saturated insulators. Brown, in his important monograph on magneto-elastic interactions [5], has given a careful investigation of the behaviour of saturated, non-dissipative and non-conductive materials, mainly under static external loading. In his book he criticizes the papers by Toupin and Tiersten. His objections especially concern the electro-magnetic foundations of these papers. Brown himself bases his theory on the model of the dipolar, magnetic continuum. In a book by Akhieser, Bar'yakhtar and Peletminskii [6], devoted to the theory of spin waves, but also containing a chapter on magneto-elastic interactions, more general aspects, e.g. magnetic dissipation and conductivity are taken into account. Both monographs [5] and [6] give some interesting applications. The book [6] also contains a good introduction to the physics of magnetic materials. Alblas has considered the magnetized body as a Cooserat-Continuum [7] and he developed this model for a more general class of problems [8]. The theory of [7] and [8] are generalized and extended in a subsequent paper [9]. Other references with respect to the literature may be found in [5] and [6].

A typical imperfection of the literature mentioned above is that the electro-magnetic foundations are not unambiguous. The only exception is the book by Brown [5], but this book is limited to the magneto-static case. For instance it is not at all obvious what has to be taken for the electric or magnetic field intensity or the electro-magnetic energy. The origin of the difficulties may be found in the lack of knowledge of the electrodynamics of moving media. In contrast to electrodynamic theory of vacuum, the subject of moving media always has been controversial.

In this paper we shall found our considerations on Chu's new formulation of electro-magnetism [10]. The basis of Chu's theory is the assumption that a material body, in rest or motion, may be considered as a set of electric and/or magnetic sources, placed in vacuum. In contrast to the conventional theory, Chu only needs two electro-magnetic vectors for the description of his field. An extensive treatment of Chu's theory with comparison with some other electro-magnetic theories and applications to relativistic and non-relativistic problems, may be found in a book by Penfield and Haus [11].

To be explicit we write down Maxwell's equations in Minkowski's formulation, in rationalized MKS units

$$\begin{aligned} \operatorname{curl} \underline{E}_M &= -\frac{\partial \underline{B}_M}{\partial t} , \\ \operatorname{curl} \underline{H}_M &= \underline{J} + \frac{\partial \underline{D}_M}{\partial t} , \\ \operatorname{div} \underline{D}_M &= \rho_{e\ell} , \\ \operatorname{div} \underline{B}_M &= 0 , \end{aligned} \qquad (1.1)$$

where $\underline{E}_M$, $\underline{H}_M$, $\underline{B}_M$ and $\underline{D}_M$ are the electric field intensity, the magnetic field intensity, the induction and the dielectric displacement, respectively. $\underline{J}$ is the current density and $\rho_{e\ell}$ the density of charge. The subscript M refers to Minkowski's definition of the quantities. In material bodies $\underline{B}_M$ and $\underline{D}_M$ depend on $\underline{H}_M$ and $\underline{E}_M$ by constitutive laws.

Chu's formulation of Maxwell's equation is

$$\begin{aligned}
\operatorname{curl} \underline{E} &= -\mu_0 \frac{\partial \underline{H}}{\partial t} - \mu_0 \rho \hat{\underline{M}} , \\
\operatorname{curl} \underline{H} &= \varepsilon_0 \frac{\partial \underline{E}}{\partial t} + \underline{J} + \rho \hat{\underline{P}} , \\
\varepsilon_0 \operatorname{div} \underline{E} &= -\operatorname{div}(\rho \underline{P}) + \rho_{e\ell} , \\
\mu_0 \operatorname{div} \underline{H} &= -\operatorname{div}(\mu_0 \rho \underline{M}) ,
\end{aligned} \tag{1.2}$$

where $\underline{E}$ and $\underline{H}$ are the electic and magnetic field intensities, respectively, $\underline{P}$ is the polarization per unit mass, $\underline{M}$ is the magnetization per unit mass, ε_0 is the permittivity and μ_0 the permeability of free space, ρ is the mass-density. We have defined the time derivative $\hat{\underline{P}}$ by

$$\rho \hat{\underline{P}} = \frac{\partial (\rho \underline{P})}{\partial t} + \operatorname{curl}(\rho \underline{P} \times \underline{v}) , \tag{1.3}$$

with $\underline{v}$ the velocity of matter. The corresponding definition for $\hat{\underline{M}}$ holds. In fact we have

$$\rho \hat{\underline{P}} = \overset{*}{\overline{\rho \underline{P}}} - \underline{v} \operatorname{div}(\rho \underline{P}) , \tag{1.4}$$

where $\overset{*}{\overline{\rho \underline{P}}}$ denotes the convective derivative [2].

The equations (1.1) and (1.2) can be brought into one-to-one correspondence if the following identifications are made

$$\begin{aligned}
\underline{D}_M &= \varepsilon_0 \underline{E} + \rho \underline{P} , \\
\underline{B}_M &= \mu_0 (\underline{H} + \rho \underline{M}) , \\
\underline{E}_M &= \underline{E} + \mu_0 \rho \underline{M} \times \underline{v} , \\
\underline{H}_M &= \underline{H} - \rho \underline{P} \times \underline{v} .
\end{aligned} \tag{1.5}$$

At this place it already becomes clear what is the origin of one of the difficulties: in different theories the same conception may refer to different quantities. However, note that in free space the formulations are identical.
A second difficulty arises from the fact that Maxwell's equations are Lorentz-invariant, while the equations of continuum mechanics are Galilei-invariant. Although it is possible to write the equations of mechanics in a Lorentz-invariant form, we shall confine ourselves to a non-relativistic treatment.
In this, Maxwell's equation must be invariant with respect to a Galilei-transformation, that is defined by

$$\underline{v} \to \underline{v} + \underline{b};\; t \to t\,,$$
$$\underline{\nabla} \to \underline{\nabla},\; \frac{\partial}{\partial t} \to \frac{\partial}{\partial t} - \underline{b}.\underline{\nabla},\; \frac{d}{dt} \to \frac{d}{dt}\,, \tag{1.6}$$

where $\underline{b}$ is a constant vector.
The relations between the electro-magnetic vectors in a system that has a constant velocity $\underline{v}$ with respect to a rest system are (in non-relativistic approximation)

$$\underline{H}^* = \underline{H} - (\underline{v} \times \varepsilon_0 \underline{E})\,,$$
$$\underline{E}^* = \underline{E} + (\underline{v} \times \mu_0 \underline{H})\,, \tag{1.7}$$
$$\underline{P}^* = \underline{P},\; \underline{M}^* = \underline{M}\,,$$

thus we may consider the starred quantities as convective ones in a reference system in constant motion $\underline{v}$.
With (1.7) we may write $(1.2)_1$ and $(1.2)_2$ in the form

$$\text{curl}\, \underline{E}^* = -\mu_0 (\overset{*}{\underline{H}} + \overset{*}{\overline{\rho \underline{M}}})\,,$$
$$\text{curl}\, \underline{H}^* = \varepsilon_0 \overset{*}{\underline{E}} + \overset{*}{\overline{\rho \underline{P}}} + \underline{J}\,, \tag{1.8}$$

if we take $\rho_{e\ell} = 0$.

Integrating (1.8) over an arbitrary surface S with boundary s, and applying Stokes' theorem, we find

$$\begin{aligned} \oint_s \underline{E}^*.\underline{ds} &= -\frac{d}{dt}\int_S \mu_0(\underline{H} + \rho\underline{M}).\underline{dS} \;, \\ \oint_s \underline{H}^*.\underline{ds} &= \frac{d}{dt}\int_S (\varepsilon_0\underline{E} + \rho\underline{P}).\underline{dS} + \int_S \underline{J}.\underline{dS} \;. \end{aligned} \tag{1.9}$$

If we assume (1.9) hold for arbitrary motions of S, (1.7) may be applied for variable values of $\underline{v}$.

In a material body the following electro-magnetic forces have to be defined: $\underline{E}$ the electric field intensity, i.e. the force on a unit charge in vacuum in a field with the prescribed sources, $\underline{E}^*$ the convective electric field intensity, $\underline{D}$ the dielectric displacement, defined according to (1.5), $\underline{G}^{(e)}$ the effective electric field intensity, i.e. the real force on a unit charge in the body. For the magnetic field we have: $\underline{H}$ the magnetic field intensity, $\underline{H}^*$ the convective magnetic field intensity, $\underline{B}$ the induction, according to $(1.5)_2$, $\underline{G}^{(m)}$ the effective magnetic field intensity. Further $\underline{f}^{(em)}$ is the body force, due to electro-magnetic actions, working on the matter. We note that none of these vectors are directly accessible to measurement.

In section 2 we derive the balance equations from a principle, first stated by Green and Rivlin [12], that may be stated as follows: The first law of thermodynamics, i.e. the energy balance equation is invariant under superposed rigid-body translations and rotations. This principle has proved its usefullness for the derivation of the known balance equations in the theory of elasticity and the theory of the Cosserat-Continuum. In our theory it yields the correct coupling terms in the balance equations, due to the electromagnetic-elastic interactions. Besides, it shows that a split of the forces into classical and electro-magnetic force is not unique and depends on our definition of "contact" forces.

In section 3 we derive the boundary and jump conditions in the non-relativistic approximation.
In section 4 we obtain some of the constitutive equations by the methods of thermodynamics.
In sections 5 and 6 we confine our discussion to the magnetization problem. First we derive the linear equations for the small extra magnetization and deformation, superposed on a state of magnetization in the body, considered as a rigid body. Our basic idea is founded upon the fact that magneto-elastic interactions occur in media in which the deformation is small. In the next section we apply our linearized theory to a few problems concerning the magneto-elastic wave propagation in an infinite space and a half-space, occupied by crystalline matter with cubic symmetry.

Balance equations

We consider a body that is placed in an external electro-magnetic field and is loaded by body and surface forces. The energy-balance equation can be written in the following form

$$\frac{d}{dt}\int_V \rho U dV + \frac{d}{dt}\frac{1}{2}\int_V \rho v_k v_k dV + \frac{d}{dt}\int_V W dV =$$

$$= \int_V \rho f_k^{(mech)} v_k dV + \int_S T_k v_k dS - \int_V G_k^{(e)} \rho \dot{P}_k dV - \int_V G_k^{(m)} \rho \dot{M}_k dV +$$

$$+ \int_S Q_k^{(e)} \dot{P}_k dS + \int_S Q_k^{(m)} \dot{M}_k dS + \int_V \rho r dV -$$

$$- \int_S h dS + \int_V G_k^{(m)} \rho Y_k dV + \int_S X dS \,. \qquad (2.1)$$

In (2.1) the energy-balance equation is applied to an arbitrary finite part of the body with volume V and bounding surface S, moving in such a way that it always consists of the same particles. Thus $\frac{d}{dt}$ denotes a material derivative. In (2.1) U is the local internal energy per unit mass and W is the electro-magnetic field energy density. The mechanical body force is denoted by $\rho \underline{f}^{(mech)}$ and the stress vector by $\underline{T}$. The quantity r denotes the heat supply per unit mass and unit time, h is the heat flux, $\underline{Q}^{(e)}$ and $\underline{Q}^{(m)}$ are the electric and magnetic surface vectors, respectively. X is an extra energy supply due to the electro-magnetic field outside of V. It contains, for instance, the Poynting energy flux and the dipole interaction energy from outside V. $\underline{Y}$ is a vector associated with the magnetic dissipation. We assume electric dissipation to be absent. We note that X is an unknown of the theory that has to be determined.

In (2.1) the interaction energy is split into two parts:

$$\int_V \rho U dV \quad \text{and} \quad \int_V W dV \,.$$

The first integral accounts for the short-range energy, e.g., the exchange coupling, the spin-orbit coupling, near magnetic and electric dipole interactions and the deformation energy. The second integral comes from the long-range interaction, e.g., the dipole-dipole interaction within V and the external field energy. The dipole-dipole interaction of particles within V with partieles outside V is included in the term

$$\int_S X dS \ .$$

Exactly stated, it is the field energy flow from outside V into V. A careful investigation of the form of W for the magneto-static case is given in [6] and [9]. Here we shall assume for W the expression from Chu's theory

$$W = \tfrac{1}{2}\varepsilon_0 E^2 + \tfrac{1}{2}\mu_0 H^2 \ . \tag{2.2}$$

The material derivative of the field energy may be calculated with the aid of (1.2) and (1.7) to be

$$\begin{aligned}\frac{d}{dt}\int_V W dV = &\int_S (\underline{H}\times\underline{E})_k n_k dS + \int_S (\tfrac{1}{2}\varepsilon_0 E^2 + \tfrac{1}{2}\mu_0 H^2) n_k v_k dS + \\ &+ \int_S (E_k^* v_k \rho P_\ell) n_\ell dS + \int_S (\mu_0 H_k^* v_k \rho M_\ell) n_\ell dS - \\ &- \int_V E_k^* J_k dV - \int_V \mu_0 (\underline{J}\times\underline{H})_k v_k dV - \\ &- \int_V E_k^* \rho \dot{P}_k dV - \int_V E_{k,\ell}^* \rho P_\ell v_k dV - \int_V \mu_0 \rho (\hat{\underline{P}}\times\underline{H})_k v_k dV - \\ &- \int_V \mu_0 H_k^* \rho \dot{M}_k dV - \int_V \mu_0 H_{k,\ell}^* \rho M_\ell v_k dV + \int_V \mu_0 \varepsilon_0 (\rho\hat{\underline{M}}\times\underline{E})_k v_k dV \ ,\end{aligned} \tag{2.3}$$

if we take $\rho_{e\ell} = 0$.

At this place we are able to simplify our equations by putting

$$X = [(\underline{H} \times \underline{E})_k + (\tfrac{1}{2}\varepsilon_0 E^2 + \tfrac{1}{2}\mu_0 H^2)v_k + (E^*_\ell v_\ell \rho P_k) + \\ + (\mu_0 H^*_\ell v_\ell \rho M_k)]n_k \ . \qquad (2.4)$$

The expression for X may easily be interpreted. It must be stressed, however, that the choice of X according to (2.4) is not unique. Another choice for X results in another separation of the electrical and the contact forces, as has been explained more extensive in [9].

Before introducing (2.3) and (2.4) into (2.1) we define the total body force $\rho\underline{f}$ by

$$\rho\underline{f} = \rho\underline{f}^{(mech)} + \rho\underline{f}^{(em)} \ . \qquad (2.5)$$

In (2.5) $\underline{f}^{(em)}$ is an unknown of the theory that has to be determined. The basis of this determination is the requirement that the body force minus the inertial force ($\rho\underline{f} - \rho\dot{\underline{v}}$) is invariant under Galilei transformations. With (2.3), (2.4) and (2.5) the balance equation (2.1) becomes

$$\int_V \rho\dot{U}dV + \int_V \rho v_k \dot{v}_k dV + \int_V (U + \tfrac{1}{2}v_k v_k)(\dot{\rho} + \rho v_{k,k})dV =$$

$$= \int_V (E^*_k - G^{(e)}_k)\rho\dot{P}_k dV + \int_V E^*_{k,\ell}\rho P_\ell v_k dV + \int_V \mu_0(\rho\hat{\underline{P}} \times \underline{H})_k v_k dV +$$

$$+ \int_V (\mu_0 H^*_k - G^{(m)}_k)\rho\dot{M}_k dV + \int_V \mu_0 H^*_{k,\ell}\rho M_\ell v_k dV -$$

$$- \int_V \mu_0\varepsilon_0(\rho\hat{\underline{M}} \times \underline{E})_k v_k dV + \int_V E^*_k J_\ell dV + \int_V \mu_0(\underline{J} \times \underline{H})_k v_k dV +$$

$$+ \int_V \rho f_k v_k dV - \int_V \rho f^{(em)}_k v_k dV + \int_S T_k v_k dS +$$

$$+ \int_S Q^{(m)}_k \dot{M}_k dS + \int_V \rho r dV - \int_S h dS + \int_V G^{(m)}_k \rho Y_k dV \ . \qquad (2.6)$$

In (2.6) we have put $Q^{(e)}$ equal to zero, because we shall not consider constitutive equations which depend on the gradient of $\underline{P}$.

We now observe the body in a frame that translates with respect to the original frame. We put

$$\underline{v} \to \underline{v} + \underline{b} , \tag{2.7}$$

where $\underline{b}$ is a constant vector. Introduction of (2.7) into (2.6) yields, by applying the condition of invariance:

$$\dot{\rho} + \rho v_{k,k} = 0 , \tag{2.8}$$

the conservation of mass, and

$$\int_V (\rho\dot{v}_k - \rho f_k)b_k dV - \int_S T_k b_k dS = 0 . \tag{2.9}$$

For the electro-magnetic body force $\rho f^{(em)}$ we find

$$\rho f_k^{(em)} = E^*_{k,\ell}\rho P_\ell + \mu_0(\rho\underline{\hat{P}} \times \underline{H}^*)_k + \mu_0(\underline{J} \times \underline{H}^*)_k -$$

$$- \mu_0\varepsilon_0(\rho\underline{\hat{M}} \times \underline{E}^*)_k + \mu_0 H^*_{k,\ell}\rho M_\ell . \tag{2.10}$$

Applying (2.9) to an infinitesimal tetrahedron we obtain

$$T_k = T_{k\ell}n_\ell \tag{2.11}$$

and from this

$$T_{k\ell,\ell} + \rho f_k = \rho\dot{v}_k . \tag{2.12}$$

Introduction of (2.5) into (2.12) yields

$$T_{k\ell,\ell} + \rho f_k^{(mech)} + \rho f_k^{(em)} = \rho\dot{v}_k . \tag{2.13}$$

In some terms of (2.10) we have replaced $\underline{H}$ and $\underline{E}$ by $\underline{H}^*$ and $\underline{E}^*$, respectively. This is consistent with the non-relativistic theory, because

$$\varepsilon_0\mu_0 = \frac{1}{c^2} , \tag{2.14}$$

where c is the velocity of light. In our calculations we neglect terms of order $(\frac{v}{c})^2$.

It is possible to write

$$\rho f_k^{(em)} = t_{k\ell,\ell} \tag{2.15}$$

where $t_{k\ell}$, the Maxwell stress, is given by

$$t_{k\ell} = E_k D_\ell - \tfrac{1}{2}\varepsilon_0 E^2 \delta_{k\ell} + H_k B_\ell - \tfrac{1}{2}\mu_0 H^2 \delta_{k\ell} + \\ + \mu_0\{(\underline{v} \times \underline{H})_k \rho P_\ell\} , \tag{2.16}$$

again in non-relativistic approximation.

The balance equation (2.6) may be simplified considerably with the aid of (2.8), (2.9) and (2.10). There results

$$\int_V \rho \dot{U} dV = \int_V (E_k^* - G_k^{(e)})\rho \dot{P}_k dV + \int_V (\mu_0 H_k^* - G_k^{(m)})\rho \dot{M}_k dV + \\ + \int_V E_k^* J_k dV + \int_V T_{k\ell} v_{k,\ell} dV + \int_S Q_k^{(m)} \dot{M}_k dS + \\ + \int_V \rho r dV - \int_S h dS + \int_V G_k^{(m)} \rho Y_k dV . \tag{2.17}$$

We have assumed the presence of magnetic dissipation, which is realized by the term

$$\int_V G_k^{(m)} \rho Y_k dV$$

in (2.17). To find a close correspondence with the physical literature, still another dissipation is introduced by writing

$$G_k^{(m)} = R_k - \eta(\dot{M}_k - e_{k\ell m}\omega_\ell M_m) , \tag{2.18}$$

where η is the coefficient of magnetic "viscosity", $e_{k\ell m}$ is the alternator and R_k is the non-dissipative part of $G_k^{(m)}$.
We have

$$\omega_\ell = \tfrac{1}{2} e_{\ell pq} v_{q,p} . \tag{2.19}$$

The introduction of (2.18) requires that the stress be split into two parts (cf. [8], [9]).

$$T_{k\ell} = \bar{T}_{k\ell} + T^{*}_{k\ell} , \tag{2.20}$$

where $\bar{T}_{k\ell}$ is the recoverable part of the stress, while $T^{*}_{k\ell}$ is the dissipative part. We introduce (2.18) and (2.20) into (2.17) and assume that the balance equation is invariant under the rigid body rotations for which hold

$$\begin{aligned} &\underline{v} \to \underline{v} + \underline{\Omega} \times \underline{r} , \\ &\dot{\underline{P}} \to \dot{\underline{P}} + \underline{\Omega} \times \underline{P} , \\ &\dot{\underline{M}} \to \dot{\underline{M}} + \underline{\Omega} \times \underline{M} , \end{aligned} \tag{2.21}$$

where $\underline{\Omega}$ is a constant vector.

If there is no magnetic dissipation, we obtain from (2.17) and (2.21)

$$\int_V \rho dV\{(R_{[k} - \mu_0 H^{*}_{[k})M_{\ell]} + (G^{(e)}_{[k} - E^{*}_{[k})P_{\ell]}\} - \int_V \bar{T}_{[k\ell]} dV - - \int_S Q^{(m)}_{[k} M_{\ell]} dS = 0 . \tag{2.22}$$

As follows from our subsequent considerations, this equation is equivalent to the requirement that the free energy be invariant with respect to rigid body rotations. If this equation also holds for the case of magnetic dissipation, we derive by the same arguments that

$$\int_V [\rho\{G^{(m)}_k Y_k + \eta(\dot{M}_k - e_{k\ell m}\omega_\ell M_m)\dot{M}_k\} + T^{*}_{k\ell} v_{k,\ell}] dV \tag{2.23}$$

must also be invariant. As under the rotation (2.21) the vector $\underline{G}^{(m)}$ transforms like (cf. [9])

$$\underline{G}^{(m)} \to \underline{G}^{(m)} - \frac{1}{\Gamma}\underline{\Omega} , \tag{2.24}$$

where Γ is the gyromagnetic ratio, we find

$$e_{pk\ell} T^{*}_{[k\ell]} + e_{pk\ell}\eta\rho(\dot{M}_k - e_{[kpq}\omega_p M_q)M_{\ell]} + \frac{\rho}{\Gamma} Y_p = 0 . \tag{2.25}$$

We introduce the mechanical angular momentum $\underline{D}$ by

$$D_p = \int_V e_{pk\ell} x_k v_\ell \rho dV \tag{2.26}$$

and find, by integrating (2.13) after multiplying with $e_{p\ell k} x_\ell$

$$\dot{D}_p = \int_V e_{p\ell k} x_\ell f_k^{(mech)} \rho dV + \int_S e_{p\ell k} x_\ell (T_k + t_{ks} n_s) dS -$$

$$- \int_V e_{p\ell k} (T_{k\ell} + t_{k\ell}) dV . \tag{2.27}$$

We define the mechanical moment $\underline{L}^{(mech)}$ by

$$L_p^{(mech)} = \int_V e_{p\ell k} x_\ell f_k^{(mech)} \rho dV + \int_S e_{p\ell k} x_\ell T_k dS \tag{2.28}$$

and the electromagnetic moment $\underline{L}^{(em)}$ by

$$L_p^{(em)} = \int_S e_{p\ell k} x_\ell t_{ks} n_s dS - \int_S e_{p\ell k} Q^{(m)}_{[\ell} M_{k]} dS . \tag{2.29}$$

Writing down the identity

$$\dot{D}_p + \frac{1}{\Gamma} \int_V \rho \dot{M}_p dV = \frac{1}{\Gamma} \int_V \rho \dot{M}_p dV + L_p^{(mech)} + L_p^{(em)} -$$

$$- \int_V e_{p\ell k} (T_{k\ell} + t_{k\ell}) dV + \int_S e_{pk\ell} Q^{(m)}_{[k} M_{\ell]} dS , \tag{2.30}$$

and making the constitutive assumption that

$$\frac{1}{\Gamma} \int_V \rho \dot{M}_p dV$$

is the magnetic part of the angular momentum, (2.30) takes the form

$$\frac{1}{\Gamma} \int_V \rho \dot{M}_p dV = - \int_V e_{pk\ell} (T_{[k\ell]} + t_{[k\ell]}) dV - \int_S e_{pk\ell} Q^{(m)}_{[k} M_{\ell]} dS, \tag{2.31}$$

if we assume the law of moment of momentum to hold for any arbitrary volume element.

From this equation we derive with (2.22), (2.20), (2.25), (2.18) and (2.16)

$$\dot{M}_p = \Gamma e_{pk\ell} M_{[k} G^{(m)}_{\ell]} + \Gamma e_{pk\ell} P_{[k} G^{(e)}_{\ell]} + Y_p \,, \tag{2.32}$$

the basic equation for the magnetic angular momentum. Note that for the derivation of (2.32) we needed an independent assumption concerning this momentum. We need another assumption for $\underline{G}^{(e)}$. We assume, following Toupin [2], the relation

$$\underline{G}^{(e)} = (\underline{A} \times \overset{*}{\overline{\rho \underline{P}}}), \tag{2.33}$$

to hold, where $\underline{A}$ is an invariant function of the deformation gradient, the polarization and the electro-magnetic field.

3. The jump and boundary conditions

Suppose that a volume V is divided by a moving surface $\Sigma(t)$ into two volumes V^- and V^+. We denote by S^- and S^+ the positions of the surface S of V which form parts of the boundaries of V^- and V^+ respectively. The remaining part of the boundary of V^- and V^+ will be furnished by the surface $\Sigma(t)$. The normal velocity of Σ along the normal to Σ is denoted by $U\underline{n}$, there $\underline{n}$ is the outward normal to Σ. We note that $U\underline{n}$ for V^- is opposite to $U\underline{n}$ for V^+.

The basic formulas for the material derivative of the integral of a function $f(x,t)$, that may have a jump over Σ are

$$\frac{d^+}{dt}\int_{V^+} f(x,t)dV = \int_{V^+}\frac{\partial f}{\partial t}\,dV + \int_{S^+} f\dot{x}_n dS - \int_{\Sigma} f^+U_n dS ,$$
$$\frac{d^-}{dt}\int_{V^-} f(x,t)dV = \int_{V^-}\frac{\partial f}{\partial t}\,dV + \int_{S^-} f\dot{x}_n dS + \int_{\Sigma} f^-U_n dS , \qquad (3.1)$$

where $\frac{d^+}{dt}$ indicates that we are to take the time derivative of the integral over a region that instantaneously coincides with V^+. U_n is the component of $U\underline{n}$ along the normal from the - side to the + side. For the whole volume V we find by adding the integrals in (3.1)

$$\frac{d}{dt}\int_V fdV = \frac{d^+}{dt}\int_{V^+} fdV + \frac{d^-}{dt}\int_{V^-} fdV =$$
$$= \int_V \frac{\partial f}{\partial t}\,dV + \oint_S f\dot{x}_n dS - \int_{\Sigma} [\![f]\!]U_n dS , \qquad (3.2)$$

where

$$[\![f]\!] = f^+ - f^- . \qquad (3.3)$$

If this equation is applied to the function $f = \rho$, we obtain

$$\int_V \frac{\partial\rho}{\partial t}\,dV + \oint_S \rho v_n dS - \int_\Sigma [\![\rho]\!] U_n dS = 0 \ . \tag{3.4}$$

Now let V approach zero at a fixed time t in such a way that it will pass, in the limit, into a part Σ_0 of the surface Σ. The volume integral in (3.4) tends to zero. But

$$\begin{aligned} \int_{S^+} \rho v_n dS &\rightarrow \int_{\Sigma_0} \rho^+ v_n^+ dS \ , \\ \int_{S^-} \rho v_n dS &\rightarrow -\int_{\Sigma_0} \rho^- v_n^- dS \ , \end{aligned} \tag{3.5}$$

where v_n^+ and v_n^- denote the normal components of the particle velocities on the + and the - sides of Σ along the normal from the - side to the + side. Hence we obtain

$$\int_{\Sigma_0} \{\rho^+(v_n^+ - U_n) - \rho^-(v_n^- - U_n)\} dS = 0 \ , \tag{3.6}$$

from which follows that

$$[\![\rho(v_n - U_n)]\!] = 0 \ . \tag{3.7}$$

This is the first jump condition. It expresses the conservation of mass. If Σ is a material surface we have

$$U_n = v_n \ , \tag{3.8}$$

and (3.7) is trivially satisfied.

A jump condition for the stresses may be obtained by taking $f = \rho v_i$ in (3.2). This equation becomes

$$\frac{d}{dt}\int_V \rho v_i dV = \int_V \frac{\partial}{\partial t}(\rho v_i)dV + \oint_S \rho v_i v_n dS - \int_\Sigma [\![\rho v_i]\!] U_n dS \,. \tag{3.9}$$

According to (2.13) and (2.15) we have

$$\frac{d}{dt}\int_V \rho v_i dV = \int_V \rho \dot{v}_i dV = \int_V \tau_{ij,j} dV + \int_V \rho f_i^{(mech)} dV =$$

$$= \oint_S \tau_{ij} n_j dS + \int_V \rho f_i^{(mech)} dV \,, \tag{3.10}$$

where the stress τ_{ij} is defined by

$$\tau_{ij} = T_{ij} + t_{ij} \,. \tag{3.11}$$

From (3.9) and (3.10) there results

$$\int_V \frac{\partial}{\partial t}(\rho v_i)dV - \int_V \rho f_i^{(mech)} dV + \oint_S \rho v_i v_n dS -$$

$$- \oint_S \tau_{ij} n_j dS - \int_\Sigma [\![\rho v_i]\!] U_n dS = 0 \,. \tag{3.12}$$

Now again we let V approach zero in the same way as before. We find

$$\int_{\Sigma_0} \{[\![\rho v_i v_n]\!] - [\![\rho v_i]\!] U_n - [\![\tau_{ij}]\!] n_j\} dS = 0 \,, \tag{3.13}$$

from which we derive, after using (3.7)

$$[\![\tau_{ij}]\!] n_j = \rho^+ (v_n^+ - U_n)[\![v_i]\!] = \rho^- (v_n^- - U_n)[\![v_i]\!] \,. \tag{3.14}$$

If U_n is the material velocity, i.e., if

$$U_n = v_n^+ = v_n^- \,, \tag{3.15}$$

(3.14) simplifies to

$$[\![\tau_{ij}]\!] n_j = [\![T_{ij}]\!] n_j + [\![t_{ij}]\!] n_j = 0 \,. \tag{3.16}$$

The application of (3.2) to the function $f = \rho M_i$ yields

$$\frac{d}{dt}\int_V \rho M_i dV = \int_V \frac{\partial}{\partial t}(\rho M_i)dV + \oint_S \rho M_i v_n dS - \int_\Sigma [\![\rho M_i]\!]U_n dS \ . \tag{3.17}$$

Integrating (2.26) we obtain

$$\frac{d}{dt}\int_V \rho M_i dV = \Gamma \int_V e_{ijk}M_{[j}G_{k]}^{(m)}\rho dV + \\ + \Gamma \int_V e_{ijk}P_{[j}G_{k]}^{(e)}\rho dV + \int_V Y_i dV \ . \tag{3.18}$$

In the next section we shall derive by independent considerations the following constitutive equations (cf. (4.11), (2.18) and (4.10))

$$G_k^{(m)} = \mu_0 H_k^* - \frac{\partial\psi}{\partial M_k} - \eta(\dot{M}_k - (\underline{\omega} \times \underline{M})_k) + \frac{1}{\rho} Q_{k\ell,\ell} \ , \tag{3.19}$$

with

$$Q_{k\ell} = \rho \frac{\partial\psi}{\partial M_{k,\alpha}} x_{\ell,\alpha} \tag{3.20}$$

and

$$G_k^{(e)} = E_k^* - \frac{\partial\psi}{\partial P_k} \ . \tag{3.21}$$

We introduce (3.18) into (3.17). With the aid of (3.19) to (3.21) we obtain

$$\int_{\Sigma_0} \{[\![\rho M_i v_n]\!] - [\![\rho M_i]\!]U_n + \Gamma e_{ijk}[\![M_{ij}Q_k^{(m)}]\!]\}dS = 0 \ , \tag{3.22}$$

from which the local equation follows after simplifying by (3.7)

$$\Gamma e_{ijk}[\![M_{[j}Q_{k]}^{(m)}]\!] + \rho^+(v_n^+ - U_n)[\![M_i]\!] = 0 \ . \tag{3.23}$$

If U_n is the material velocity we have

$$[\![\underline{M} \times \underline{Q}^{(m)}]\!] = 0 \ . \tag{3.24}$$

The application of (3.16) requires the calculation of the jump

$$[\![t_{ij}]\!]n_j = [\![t_i]\!] = \overset{+}{t}_{ij}n_j - \overset{-}{t}_{ij}n_j \qquad (3.25)$$

in the Maxwell stress. From (1.2) and (1.9) we derive

$$[\![\varepsilon_0\underline{E} + \rho\underline{P}]\!].\underline{n} = 0 \ , \qquad (3.26)$$

$$[\![\underline{H} + \rho\underline{M}]\!].\underline{n} = 0 \ , \qquad (3.27)$$

$$\underline{n} \times [\![\underline{H}]\!] + U_n\varepsilon_0[\![\underline{E}]\!] = 0 \ , \qquad (3.28)$$

$$\underline{n} \times [\![\underline{E}]\!] - U_n\mu_0[\![\underline{H}]\!] = 0 \ . \qquad (3.29)$$

For a material surface we see that the tangential components of $\underline{E}^*$ and $\underline{H}^*$, E_t^* and H_t^*, respectively, are continuous. The normal components of $\underline{D}$ and $\underline{B}$ are always continuous.

With (3.26) to (3.29) we now calculate (3.25) for a material surface. We find

$$\overset{+}{t}_k - \overset{-}{t}_k = \overset{+}{E}_k\overset{+}{D}_n - \overset{-}{E}_k\overset{-}{D}_n - \tfrac{1}{2}\varepsilon_0(E^{+2} - E^{-2})n_k +$$

$$+ \overset{+}{H}_k\overset{+}{B}_n - \overset{-}{H}_k\overset{-}{B}_n - \tfrac{1}{2}\mu_0(H^{+2} - H^{-2})n_k +$$

$$+ \mu_0(\underline{v} \times \underline{H}^+)_k\rho^+\overset{+}{P}_\ell n_\ell - \mu_0(\underline{v} \times \underline{H}^-)_k\rho^-\overset{-}{P}_\ell n_\ell =$$

$$= -\frac{1}{2\varepsilon_0}[\![\rho^2P_n^2]\!]n_k - \tfrac{1}{2}\mu_0[\![\rho^2M_n^2]\!]n_k \ . \qquad (3.30)$$

From these jump conditions the boundary conditions may easily be formulated, because the surface S of the body is a material surface. We have

$$T_k^{(0)} = T_k - \frac{1}{2\varepsilon_0}\rho^2P_n^2n_k - \frac{\mu_0}{2}\rho^2M_n^2n_k \ , \quad \text{at } S \ , \qquad (3.31)$$

within the approximation of the non-relativistic theory. In (3.31) $T_k^{(0)}$ is the prescribed stress vector. From (3.24) we conclude that

$$\underline{M} \times \underline{Q}^m = 0, \text{ at } S \ . \qquad (3.32)$$

From energy considerations (cf. [9]) we obtain that at the bounding surface

$$\int_S Q_k^{(m)}\dot{M}_k dS = 0 \ , \qquad (3.33)$$

that may be written by (2.26) in the form

$$\int_S [\Gamma e_{kpq} Q_k^{(m)} M_{[p} G_{q]}^{(m)} + \Gamma e_{kpq} Q_k^{(m)} P_{[k} G_{q]}^{(e)} + Q_k^{(m)} Y_k] dS = 0 \ . \quad (3.34)$$

In general (3.34) and (3.32) are only consistent if

$$\underline{Q}^{(m)} = 0, \text{ at } S \ . \quad (3.35)$$

Only if $\underline{P} = \underline{Y} = 0$, (3.34) implies (3.32).

However, it appears that for states of constant magnetization, we always have to take (3.35) as boundary condition.

4. The constitutive equations

Constitutive equations are required for $\bar{T}_{(k\ell)}$, $\bar{T}_{[k\ell]}$, $T^*_{(k\ell)}$, Y_k, R_k, $Q_k^{(m)}$, J_k and h_k. The antisymmetric tensor $T^*_{[k\ell]}$ is determined by (2.25), while $G_k^{(e)}$ is given by (2.27).

Part of the constitutive equations may be obtained by thermodynamic considerations. We may introduce the specific entropy η by the Clausius-Duhem inequality

$$\frac{d}{dt}\int_V \rho\eta dV - \int_V \frac{\rho r}{\theta} dV + \int_S \frac{h}{\theta} dS \geq 0 , \tag{4.1}$$

where θ is the local temperature. Applying (4.1) to an infinitesimal tetrahedron we obtain

$$h = h_k n_k , \tag{4.2}$$

from which a local inequality follows

$$\rho\theta\dot{\eta} - \rho r + h_{k,k} - \frac{h_k \theta_{,k}}{\theta} \geq 0 . \tag{4.3}$$

We use this inequality for the elimination of

$$\int_S h dS - \int_V \rho r dV$$

from (2.17). After the introduction of the Helmholtz free energy ψ

$$\psi = U - \theta\eta , \tag{4.4}$$

we obtain from (4.3) and (2.17)

$$\begin{aligned} &\int_V [\rho\{\dot{\psi} + \eta\dot{\theta} + (G_k^{(e)} - E_k^*)\dot{P}_k + (G_k^{(m)} - \mu_0 H_k^*)\dot{M}_k\} - \\ &- T_{k\ell} v_{k,\ell}]dV - \int_V E_k^* J_k dV - \int_V G_k^{(m)} Y_k \rho dV - \\ &- \int_S Q_k^{(m)} \dot{M}_k dS + \int_V \frac{h_k \theta_{,k}}{\theta} dV \leq 0 . \end{aligned} \tag{4.5}$$

The function ψ is assumed to be a function of θ, $x_{k,\alpha}$, P_k, M_k and $M_{k,\alpha}$:

$$\psi = \psi(\theta, x_{k,\alpha}, P_k, M_k, M_{k,\alpha}) \ . \tag{4.6}$$

From (4.6) we derive

$$\dot{\psi} = \frac{\partial\psi}{\partial\theta}\dot{\theta} + \frac{\partial\psi}{\partial x_{k,\alpha}}\dot{x}_{k,\alpha} + \frac{\partial\psi}{\partial P_k}\dot{P}_k + \frac{\partial\psi}{\partial M_k}\dot{M}_k + \frac{\partial\psi}{\partial M_{k,\alpha}}\dot{M}_{k,\alpha} \ . \tag{4.7}$$

If we introduce (4.7) into (4.5) and then apply the usual thermodynamic arguments we find

$$\eta = -\frac{\partial\psi}{\partial\theta} \ , \tag{4.8}$$

$$\bar{T}_{k\ell} = \rho\frac{\partial\psi}{\partial x_{k,\alpha}} x_{\ell,\alpha} \ , \tag{4.9}$$

$$G_k^{(e)} = E_k^* - \frac{\partial\psi}{\partial P_k} \ , \tag{4.10}$$

$$R_k = \mu_0 H_k^* - \frac{\partial\psi}{\partial M_k} + \frac{1}{\rho} Q_{k\ell,\ell} \tag{4.11}$$

$$Q_k^{(m)} = Q_{k\ell} n_\ell \ . \tag{4.12}$$

The inequality (4.5), with (4.8) to (4.12) simplifies to

$$\int_V [T^*_{k\ell} v_{k,\ell} + E^*_k J_k + G^{(m)}_k Y_k \rho + \eta\rho(\dot{M}_k - (\underline{\omega} \times \underline{M})_k)\dot{M}_k - \\ - \frac{h_k \theta_{,k}}{\theta}]\, dV \geq 0 \ , \tag{4.13}$$

from which we conclude that

$$-\frac{h_k \theta_{,k}}{\theta} \geq 0 \ , \tag{4.14}$$

$$T^*_{(k\ell)} v_{(k,\ell)} \geq 0 \ , \tag{4.15}$$

$$\eta\rho(\dot{M}_k - (\underline{\omega} \times \underline{M})_k)(\dot{M}_k - (\underline{\omega} \times \underline{M})_k) \geq 0 \ , \tag{4.16}$$

$$E_k^* J_k \geq 0 \; , \tag{4.17}$$

$$(G_k^{(m)} + \frac{\omega_k}{\Gamma}) \rho Y_k \geq 0 \; . \tag{4.18}$$

For the derivations of (4.15) and (4.18) we used (2.25).

We shall assume linear equations for the quantities that are of minor importance for our theory and take

$$h_k = -\kappa \theta_{,k} \; , \; \kappa > 0 \; , \tag{4.19}$$

$$T^*_{(k\ell)} = \mu_1 \delta_{k\ell} v_{p,p} + 2\mu_2 v_{(k,\ell)} , \; \mu_1 + \frac{2}{3}\mu_2 \geq 0 \; , \tag{4.20}$$

$$J_k = \sigma_{k\ell} E^*_\ell , \; \sigma_{k\ell} x_k x_\ell \geq 0 \; , \tag{4.21}$$

$$Y_k = \chi_{k\ell} (G_\ell + \frac{\omega_\ell}{\Gamma}) , \; \chi_{k\ell} x_k x_\ell \geq 0 \; . \tag{4.22}$$

To proceed we note that, if ψ has to remain invariant if the body undergoes a rigid rotation, it must have the form

$$\psi = \psi(\theta, A_{\alpha\beta}, B_\alpha, C_{\alpha\beta}, D_\alpha) \; , \tag{4.23}$$

where

$$A_{\alpha\beta} = x_{k,\alpha} M_{k,\beta} \; , \tag{4.24}$$

$$B_\alpha = x_{k,\alpha} M_k \; , \tag{4.25}$$

$$C_{\alpha\beta} = x_{k,\alpha} x_{k,\beta} \; , \tag{4.26}$$

$$D_\alpha = x_{k,\alpha} P_k \; . \tag{4.27}$$

From (4.24) to (4.27) we derive

$$\frac{\partial A_{\gamma\delta}}{\partial x_{k,\alpha}} = \delta_{\gamma\alpha} M_{k,\delta} \; , \tag{4.28}$$

$$\frac{\partial B_\gamma}{\partial x_{k,\alpha}} = \delta_{\gamma\alpha} M_k \; , \tag{4.29}$$

$$\frac{\partial C_{\gamma\delta}}{\partial x_{k,\alpha}} = \delta_{\gamma\alpha} x_{k,\delta} + \delta_{\delta\alpha} x_{k,\gamma} , \tag{4.30}$$

$$\frac{\partial D_{\gamma}}{\partial x_{k,\alpha}} = \delta_{\gamma\alpha} P_k . \tag{4.31}$$

Introduction of the equalities into the constitutive equations (4.9) to (4.12) results in

$$\bar{T}_{k\ell} = \rho x_{\ell,\alpha} [\frac{\partial\psi}{\partial A_{\alpha\beta}} M_{k,\beta} + \frac{\partial\psi}{\partial B_{\alpha}} M_k + 2 \frac{\partial\psi}{\partial C_{\alpha\beta}} x_{k,\beta} + \frac{\partial\psi}{\partial D_{\alpha}} P_k] , \tag{4.32}$$

$$G_k^{(e)} = E_k^* - \frac{\partial\psi}{\partial D_{\alpha}} x_{k,\alpha} , \tag{4.33}$$

$$R_k = \mu_0 H_k^* - \frac{\partial\psi}{\partial B_{\alpha}} x_{k,\alpha} + \frac{1}{\rho} (\rho \frac{\partial\psi}{\partial A_{\gamma\alpha}} x_{k,\gamma} x_{\ell,\alpha})_{,\ell} , \tag{4.34}$$

$$Q_k^{(m)} = \rho \frac{\partial\psi}{\partial A_{\alpha\beta}} x_{k,\alpha} x_{\ell,\beta} n_{\ell} . \tag{4.35}$$

5. Linearization of the magnetic problem

In this and the subsequent section we confine our discussions to the magnetic problem, i.e., we take $\underline{P} = 0$. We assume that the deformations are infinitesimately small and that the pure (rigid) magnetic problem has been solved. This means that we are looking for solutions that are perturbations of the magnetic equations due to the elastic deformations.

We set

$$x_k = X_k + u_k \ , \tag{5.1}$$

$$x_{k,\alpha} = X_{k,\alpha} + u_{k,\alpha} = \delta_{k\alpha} + u_{k,\alpha} \ , \tag{5.2}$$

$$M_k = \tilde{M}_k + m_k \ , \tag{5.3}$$

$$M_{k,\alpha} = \tilde{M}_{k,\alpha} + m_{k,\alpha} \ , \tag{5.4}$$

where $\tilde{M}_k$ is the magnetization of the rigid body.

Up to the first order we have

$$A_{\alpha\beta} = \tilde{M}_{\alpha,\beta} + m_{\alpha,\beta} + u_{k,\alpha}\tilde{M}_{k,\beta} \ , \tag{5.5}$$

$$B_\alpha = \tilde{M}_\alpha + m_\alpha + u_{k,\alpha}\tilde{M}_k \ , \tag{5.6}$$

$$C_{\alpha\beta} = \delta_{\alpha\beta} + u_{\alpha,\beta} + u_{\beta,\alpha} \ . \tag{5.7}$$

For the density we take

$$\rho = \rho_0(1 - u_{k,k}) \ , \tag{5.8}$$

with ρ_0 the density of the rigid body.

We develop the functions $\frac{\partial\psi}{\partial A_{\alpha\beta}}$, $\frac{\partial\psi}{\partial B_\alpha}$, $\frac{\partial\psi}{\partial C_{\alpha\beta}}$ in the following form

$$\frac{\partial\psi}{\partial A_{\alpha\beta}} = \left(\frac{\partial\psi}{\partial A_{\alpha\beta}}\right)_0 + \left(\frac{\partial^2\psi}{\partial A_{\alpha\beta}\partial A_{\gamma\delta}}\right)_0(m_{\gamma,\delta} + u_{k,\gamma}\tilde{M}_{k,\delta}) + \left(\frac{\partial^2\psi}{\partial A_{\alpha\beta}\partial B_\gamma}\right)_0(m_\gamma + u_{k,\gamma}\tilde{M}_k) + \left(\frac{\partial^2\psi}{\partial A_{\alpha\beta}\partial C_{\gamma\delta}}\right)_0(u_{\gamma,\delta} + u_{\delta,\gamma}) \tag{5.9}$$

$$\frac{\partial\psi}{\partial B_\alpha} = \left(\frac{\partial\psi}{\partial B_\alpha}\right)_0 + \left(\frac{\partial^2\psi}{\partial B_\alpha \partial A_{\beta\gamma}}\right)_0 (m_{\beta,\gamma} + u_{k,\beta}\tilde{M}_{k,\gamma}) +$$

$$+ \left(\frac{\partial^2\psi}{\partial B_\alpha \partial B_\beta}\right)_0 (m_\beta + u_{k,\beta}\tilde{M}_k) + 2\left(\frac{\partial^2\psi}{\partial B_\alpha \partial C_{\gamma\delta}}\right)_0 u_{\beta,\gamma} , \tag{5.10}$$

$$\frac{\partial\psi}{\partial C_{\alpha\beta}} = \left(\frac{\partial\psi}{\partial C_{\alpha\beta}}\right)_0 + \left(\frac{\partial^2\psi}{\partial C_{\alpha\beta} \partial A_{\gamma\delta}}\right)_0 (m_{\gamma,\delta} + u_{k,\gamma}\tilde{M}_{k,\delta}) +$$

$$+ \left(\frac{\partial^2\psi}{\partial C_{\alpha\beta} \partial B_\gamma}\right)_0 (m_\gamma + u_{k,\gamma}\tilde{M}_k) + 2\left(\frac{\partial^2\psi}{\partial C_{\alpha\beta} \partial C_{\gamma\delta}}\right)_0 u_{\gamma,\delta} , \tag{5.11}$$

where $(\;)_0$ denotes the values of the functions for $m_k = u_k = 0$. From (4.32) and (5.5) to (5.11) we find

$$\bar{T}_{k\ell} = \tilde{\bar{T}}_{k\ell} + \bar{T}'_{k\ell} \tag{5.12}$$

with

$$\tilde{\bar{T}}_{k\ell} = \rho_0\left[\tilde{M}_{k,\delta}\left(\frac{\partial\psi}{\partial A_{\ell\delta}}\right)_0 + \tilde{M}_k\left(\frac{\partial\psi}{\partial B_\ell}\right)_0 + 2\left(\frac{\partial\psi}{\partial C_{k\ell}}\right)_0\right] \tag{5.13}$$

and

$$\bar{T}'_{k\ell} = a_{k\ell p} m_p + b_{k\ell pq} m_{p,q} + c_{k\ell pq} u_{p,q} \tag{5.14}$$

with

$$a_{k\ell p} = \rho_0\left[\tilde{M}_{k,\delta}\left(\frac{\partial^2\psi}{\partial A_{\ell\delta}\partial B_p}\right)_0 + \left(\frac{\partial\psi}{\partial B_\ell}\right)_0 \delta_{kp} +\right.$$

$$\left. + \tilde{M}_k\left(\frac{\partial^2\psi}{\partial B_\ell \partial B_p}\right)_0 + 2\left(\frac{\partial^2\psi}{\partial C_{k\ell}\partial B_p}\right)_0\right] , \tag{5.15}$$

$$b_{k\ell pq} = \rho_0\left[\left(\frac{\partial\psi}{\partial A_{\ell q}}\right)_0 \delta_{kp} + \tilde{M}_{k,\delta}\left(\frac{\partial^2\psi}{\partial A_{\ell\delta}\partial A_{pq}}\right)_0 +\right.$$

$$\left. + \tilde{M}_k\left(\frac{\partial^2\psi}{\partial B_\ell \partial A_{pq}}\right)_0 + 2\left(\frac{\partial^2\psi}{\partial C_{k\ell}\partial A_{pq}}\right)_0\right] , \tag{5.16}$$

$$c_{k\ell pq} = \rho_0[\tilde{M}_{k,\delta}\tilde{M}_{p,\beta}(\frac{\partial^2\psi}{\partial A_{\ell\delta}\partial A_{q\beta}})_0 + \tilde{M}_{k,\delta}\tilde{M}_p(\frac{\partial^2\psi}{\partial A_{\ell\delta}\partial B_q})_0 +$$

$$+ 2\tilde{M}_{k,\delta}(\frac{\partial^2\psi}{\partial A_{\ell\delta}\partial C_{pq}})_0 + \tilde{M}_k\tilde{M}_{p,\gamma}(\frac{\partial^2\psi}{\partial B_\ell\partial A_{q\gamma}})_0 +$$

$$+ \tilde{M}_k\tilde{M}_p(\frac{\partial^2\psi}{\partial B_\ell\partial B_q})_0 + 2\tilde{M}_k(\frac{\partial^2\psi}{\partial B_\ell\partial C_{pq}})_0 +$$

$$+ 2\tilde{M}_{p,\delta}(\frac{\partial^2\psi}{\partial C_{k\ell}\partial A_{q\delta}})_0 + 2\tilde{M}_p(\frac{\partial^2\psi}{\partial B_p\partial C_{k\ell}})_0 +$$

$$+ 4(\frac{\partial^2\psi}{\partial C_{k\ell}\partial C_{pq}})_0 + 2(\frac{\partial\psi}{\partial C_{\ell q}})_0\delta_{kp} + \tilde{M}_{k,\delta}(\frac{\partial\psi}{\partial A_{q\delta}})_0\delta_{p\ell} +$$

$$+ \tilde{M}_k(\frac{\partial\psi}{\partial B_q})_0\delta_{p\ell} + 2(\frac{\partial\psi}{\partial C_{qk}})_0\delta_{p\ell} - \tilde{M}_{k,\delta}(\frac{\partial\psi}{\partial A_{\ell\delta}})_0\delta_{pq} -$$

$$- \tilde{M}_k(\frac{\partial\psi}{\partial B_\ell})_0\delta_{pq} - 2(\frac{\partial\psi}{\partial C_{k\ell}})_0\delta_{pq}] . \tag{5.17}$$

To the same order we expand the function R_k, using (4.34) and (5.5) to (5.11)

$$R_k = \tilde{R}_k + R'_k , \tag{5.18}$$

with

$$\tilde{R}_k = \mu_0 H_k - (\frac{\partial\psi}{\partial B_k})_0 + (\frac{\partial\psi}{\partial A_{k\ell}})_{0,\ell} \tag{5.19}$$

and

$$R'_k = \mu_0 h^*_k + d_{kp}m_p + e_{kpq}m_{p,q} + f_{kpq}u_{p,q} +$$

$$+ g_{kpqr}m_{p,qr} + h_{kpqr}u_{p,qr} , \tag{5.20}$$

with

$$d_{kp} = -\left(\frac{\partial^2\psi}{\partial B_k \partial B_p}\right)_0 + \left(\frac{\partial^2\psi}{\partial A_{k\ell} \partial B_p}\right)_{0,\ell} , \tag{5.21}$$

$$e_{kpq} = -\left(\frac{\partial^2\psi}{\partial B_k \partial A_{pq}}\right)_0 + \left(\frac{\partial^2\psi}{\partial A_{k\ell} \partial A_{pq}}\right)_{0,\ell} + \left(\frac{\partial^2\psi}{\partial A_{k\ell} \partial B_p}\right)_0 \delta_{q\ell} , \tag{5.22}$$ *)

$$\begin{aligned} f_{kpq} = & -\left(\frac{\partial\psi}{\partial B_q}\right)_0 \delta_{kp} - \left(\frac{\partial^2\psi}{\partial B_k \partial A_{q\gamma}}\right)_0 \tilde{M}_{p,\gamma} - \\ & - \left(\frac{\partial^2\psi}{\partial B_k \partial B_q}\right)_0 \tilde{M}_p - 2\left(\frac{\partial^2\psi}{\partial B_k \partial C_{pq}}\right)_0 + \left(\frac{\partial\psi}{\partial A_{q\ell}}\right)_{0,\ell} \delta_{pk} + \\ & + \left(\frac{\partial\psi}{\partial A_{kq}}\right)_{0,\ell} \delta_{p\ell} + \left\{\left(\frac{\partial^2\psi}{\partial A_{k\ell} \partial A_{qt}}\right)_0 \tilde{M}_{p,t}\right\}_{,\ell} + \\ & + \left\{\left(\frac{\partial^2\psi}{\partial A_{k\ell} \partial B_q}\right)_0 \tilde{M}_p\right\}_{,\ell} + 2\left(\frac{\partial^2\psi}{\partial A_{k\ell} \partial C_{pq}}\right)_{0,\ell} , \end{aligned} \tag{5.23}$$

$$g_{kpqr} = \left(\frac{\partial^2\psi}{\partial A_{kr} \partial A_{pq}}\right)_0 , \tag{5.24}$$

$$\begin{aligned} h_{kpqr} = & -\left(\frac{\partial\psi}{\partial A_{kr}}\right)_0 \delta_{pq} + \left(\frac{\partial\psi}{\partial A_{qr}}\right)_0 \delta_{pk} + \left(\frac{\partial\psi}{\partial A_{kq}}\right)_0 \delta_{pr} + \\ & + \left(\frac{\partial^2\psi}{\partial A_{kr} \partial A_{qt}}\right) \tilde{M}_{p,t} + \left(\frac{\partial^2\psi}{\partial A_{kr} \partial B_q}\right)_0 \tilde{M}_p + 2\left(\frac{\partial^2\psi}{\partial A_{kr} \partial C_{pq}}\right)_0 . \end{aligned} \tag{5.25}$$

In (5.20) the magnetic field intensity h_k^* is defined by

$$h_k^* = h_k - (\underline{v} \times \varepsilon_0 \underline{E})_k , \tag{5.26}$$

with h_k the external and internal extra magnetic field intensity.

---

*) This coefficient must not be mistaken for the alternator e_{kpq}.

We now expand the function $Q_k^{(m)}$ according to (4.35) by using (5.5) to (5.11)

$$Q_k^{(m)} = \tilde{Q}_k + Q'_k , \tag{5.27}$$

with

$$\tilde{Q}_k = \rho_0 \left(\frac{\partial\psi}{\partial A_{k\ell}}\right)_0 n_\ell \tag{5.28}$$

$$Q'_k = s_{kp} m_p + s_{kpq} m_{p,q} + t_{kpq} u_{p,q} , \tag{5.29}$$

with

$$s_{kp} = \rho_0 \left(\frac{\partial^2\psi}{\partial A_{k\ell}\partial B_p}\right)_0 n_\ell , \tag{5.30}$$

$$s_{kpq} = \rho_0 \left(\frac{\partial^2\psi}{\partial A_{k\ell}\partial A_{pq}}\right)_0 n_\ell , \tag{5.31}$$

$$t_{kpq} = \rho_0 n_\ell \Big[-\left(\frac{\partial\psi}{\partial A_{k\ell}}\right)_0 \delta_{pq} + \left(\frac{\partial\psi}{\partial A_{q\ell}}\right)_0 \delta_{kp} + \left(\frac{\partial\psi}{\partial A_{kq}}\right)_0 \delta_{p\ell} + \left(\frac{\partial^2\psi}{\partial A_{k\ell}\partial A_{q\beta}}\right)_0 \tilde{M}_{p,\beta} + \left(\frac{\partial^2\psi}{\partial A_{k\ell}\partial B_q}\right)_0 \tilde{M}_p + 2\left(\frac{\partial^2\psi}{\partial A_{k\ell}\partial C_{pq}}\right)_0\Big] . \tag{5.32}$$

To be able to perform some calculations we shall have to represent the function ψ explicitly. Physical considerations lead to the assumption that we may write

$$\psi = \psi_1(A_{\alpha\beta}, C_{\alpha\beta}) + \psi_2(B_\alpha, C_{\alpha\beta}) . \tag{5.33}$$

It is sufficient to expand (5.33) in the following forms

$$\psi_1 = p^{(13)}_{\alpha\beta\gamma\delta} A_{\alpha\beta} A_{\gamma\delta} + p^{(16)}_{\alpha\beta\gamma\delta\epsilon\phi} A_{\alpha\beta} A_{\gamma\delta} C_{\epsilon\phi} , \tag{5.34}$$

$$\psi_2 = q^{(20)} + q^{(21)}_{\alpha\beta} C_{\alpha\beta} + q^{(22)}_{\alpha\beta\gamma\delta} C_{\alpha\beta} C_{\gamma\delta} , \tag{5.35}$$

where the coefficients p and q satisfy self evident symmetry relations, the p's are constants, while the q's depend only on $\underline{B}$.

With (5.34) and (5.35) we may calculate the coefficients in the constitutive equations. If we assume the basic state to be a state of constant magnetization, the following coefficients disappear:

$$b_{k\ell pq} = e_{kpq} = h_{kpqr} = s_{kp} = t_{kpq} = 0 \; . \tag{5.36}$$

With the abbreviations

$$q^{(1)} = q^{(20)} + q_{ss}^{(21)} + q_{sstt}^{(22)} , \tag{5.37}$$

$$q_{k\ell}^{(2)} = q_{k\ell}^{(21)} + 2q_{k\ell ss}^{(22)} , \tag{5.38}$$

we find for this case

$$a_{k\ell p} = \rho_0 \left(\frac{\partial q^{(1)}}{\partial \tilde{M}_\ell} \delta_{pk} + \tilde{M}_k \frac{\partial^2 q^{(1)}}{\partial \tilde{M}_\ell \partial \tilde{M}_p} + 2 \frac{\partial q_{k\ell}^{(2)}}{\partial \tilde{M}_p} \right) , \tag{5.39}$$

$$\begin{aligned} c_{k\ell pq} = \rho_0 \Big[& \tilde{M}_k \tilde{M}_p \frac{\partial^2 q^{(1)}}{\partial \tilde{M}_\ell \partial \tilde{M}_q} + 2\tilde{M}_k \frac{\partial q_{pq}^{(2)}}{\partial \tilde{M}_\ell} + \\ & + 2\tilde{M}_p \frac{\partial q_{k\ell}^{(2)}}{\partial \tilde{M}_q} + 8q_{k\ell pq}^{(2)} + 2\delta_{kp} q_{\ell q}^{(2)} + \tilde{M}_k \frac{\partial q^{(1)}}{\partial \tilde{M}_q} \delta_{p\ell} + \\ & + 2\delta_{p\ell} q_{qk}^{(2)} - \tilde{M}_k \delta_{pq} \frac{\partial q^{(1)}}{\partial \tilde{M}_\ell} - 2\delta_{pq} q_{k\ell}^{(2)} \Big] \end{aligned} \tag{5.40}$$

$$d_{kp} = - \frac{\partial^2 q^{(1)}}{\partial \tilde{M}_k \partial \tilde{M}_p} , \tag{5.41}$$

$$f_{kpq} = -\delta_{kp} \frac{\partial q^{(1)}}{\partial \tilde{M}_q} - \tilde{M}_p \frac{\partial^2 q^{(1)}}{\partial \tilde{M}_k \partial \tilde{M}_q} - 2 \frac{\partial q_{pq}^{(2)}}{\partial \tilde{M}_k} , \tag{5.42}$$

$$g_{kpqr} = 2p_{krpq}^{(13)} + 2p_{krpqss}^{(16)} , \tag{5.43}$$

$$s_{kpq} = \rho_0 n_\ell g_{kpq\ell} \; . \tag{5.44}$$

We now further restrict our discussion to the consideration of magnetic phenomena in a body of crystalline material of cubic symmetry. The free energy of a cubic crystal, to a first approximation, may be written as

$$\psi_1 = \tfrac{1}{2}\alpha_1 M_{k,\ell} M_{k,\ell} + \tfrac{1}{2}\alpha_2 u_{s,s} M_{k,\ell} M_{k,\ell} \ , \tag{5.45}$$

$$\psi_2 = q^{(1)} + 2q^{(2)}_{\alpha\beta} u_{\alpha,\beta} + 4q^{(22)}_{\alpha\beta\gamma\delta} u_{\alpha,\beta} u_{\gamma,\delta} \ . \tag{5.46}$$

In (5.45) α_1 and α_2 are exchange parameters.
We have the following expansions

$$q^{(1)} = g^{(1)}_{k\ell mn} \tilde{M}_k \tilde{M}_\ell \tilde{M}_m \tilde{M}_n + \ldots \ , \tag{5.47}$$

$$q^{(2)}_{\alpha\beta} = g^{(2)}_{\alpha\beta k\ell} \tilde{M}_k \tilde{M}_\ell + \ldots \ , \tag{5.48}$$

that are broken off at the lower order terms. We assume $q^{(22)}_{\alpha\beta\gamma\delta}$ to be independent of $\underline{\tilde{M}}$. Cubic symmetry requires, if we take the coordinate axes along the edges of the cube

$$g^{(1)}_{1122} = g^{(1)}_{2233} = g^{(1)}_{3311} = g^{(1)}_{2211} = g^{(1)}_{3322} = g^{(1)}_{1133} = \frac{k_1}{12} \ , \tag{5.49}$$

while the other $g^{(1)}_{k\ell mn}$ are equal to zero,

$$g^{(2)}_{1111} = g^{(2)}_{2222} = g^{(2)}_{3333} = \frac{k_2}{4} \ , \tag{5.50}$$

$$g^{(2)}_{k\ell k\ell} = g^{(2)}_{\ell k\ell k} = g^{(2)}_{k\ell\ell k} = g^{(2)}_{\ell kk\ell} = \frac{k_3}{4} \ , \tag{5.51}$$

$(\ell \neq k,\ k,\ell$ not summed$)$,

and the other $g^{(2)}_{\alpha\beta k\ell}$ equal to zero. In (5.49), (5.50) and (5.51) k_1, k_2 and k_3 are constants. For the elasticity constants $q^{(22)}_{\alpha\beta\gamma\delta}$ we have

$$q^{(22)}_{1111} = q^{(22)}_{2222} = q^{(22)}_{3333} = \frac{1}{8} c_1 \ , \tag{5.52}$$

$$q^{(22)}_{1122} = q^{(22)}_{2233} = q^{(22)}_{3311} = q^{(22)}_{2211} = q^{(22)}_{3322} = q^{(22)}_{1133} = \frac{c_2}{8} \ , \tag{5.53}$$

$$q^{(22)}_{pqpq} = q^{(22)}_{pqqp} = q^{(22)}_{qppq} = q^{(22)}_{qpqp} = \frac{c_3}{8} , \tag{5.54}$$

$(p \neq q$, p,q not summed) ,

while the other $q^{(22)}_{\alpha\beta\gamma\delta}$ are equal to zero.

We use the expansions (5.47) and (5.48) for the calculations of the coefficients. We obtain for the case of constant magnetization

$$a_{k\ell p} = \rho_0 (4g^{(1)}_{s\ell mn}\tilde{M}_s\tilde{M}_m\tilde{M}_n\delta_{pk} + 12g^{(1)}_{\ell pst}\tilde{M}_k\tilde{M}_s\tilde{M}_t + 4g^{(2)}_{k\ell pt}\tilde{M}_t) , \tag{5.55}$$

$$\begin{aligned} c_{k\ell pq} = \rho_0 [& 8q^{(22)}_{k\ell pq} + 12g^{(1)}_{\ell qmn}\tilde{M}_k\tilde{M}_p\tilde{M}_m\tilde{M}_n + 4g^{(2)}_{pq\ell t}\tilde{M}_k\tilde{M}_t + 4g^{(2)}_{k\ell qt}\tilde{M}_p\tilde{M}_t + \\ & + 2g^{(2)}_{\ell qmn}\tilde{M}_m\tilde{M}_n\delta_{kp} + 4\delta_{p\ell}g^{(1)}_{qmns}\tilde{M}_k\tilde{M}_m\tilde{M}_n\tilde{M}_s + 2\delta_{p\ell}g^{(2)}_{qkst}\tilde{M}_s\tilde{M}_t - \\ & - 4\delta_{pq}g^{(1)}_{\ell mns}\tilde{M}_k\tilde{M}_m\tilde{M}_n\tilde{M}_s - 2\delta_{pq}g^{(2)}_{k\ell mn}\tilde{M}_m\tilde{M}_n] , \end{aligned} \tag{5.56}$$

$$d_{kp} = -12g^{(1)}_{kpmn}\tilde{M}_m\tilde{M}_n , \tag{5.57}$$

$$f_{kpq} = -4\delta_{kp}g^{(1)}_{qmns}\tilde{M}_m\tilde{M}_n\tilde{M}_s - 12g^{(1)}_{kqmn}\tilde{M}_p\tilde{M}_m\tilde{M}_n - 4g^{(2)}_{pqkm}\tilde{M}_m , \tag{5.58}$$

$$g_{kpqr} = \alpha_1\delta_{kp}\delta_{qr} , \tag{5.59}$$

$$s_{kpq} = \rho_0\alpha_1\delta_{kp}n_q . \tag{5.60}$$

From (5.56) it may easily be seen that the coefficients $c_{k\ell pq}$ do not satisfy the symmetry relations (5.52) to (5.54) which hold for $q_{k\ell pq}$. As could be expected the cubic symmetry is disturbed by the appearance of a magnetization vector.

For future reference we write down the formulas for $T'_{k\ell}$, R'_k and Q'_k if the material has cubic symmetry and is magnetized to saturation along one of the edges of the cube, which we take to be the z-axis. We find from (5.55) to (5.60) and (5.49) to (5.51)

$$
\begin{aligned}
T'_{xx} &= c_{xxpq}u_{p,q} \,, \\
T'_{xy} &= c_{xypq}u_{p,q} \,, \\
T'_{xz} &= k_3\rho_0 M_s m_x + c_{xzpq}u_{p,q} \,, \\
T'_{yx} &= c_{yxpq}u_{p,q} \,, \\
T'_{yy} &= c_{yypq}u_{p,q} \,, \\
T'_{yz} &= k_3\rho_0 M_s m_y + c_{yzpq}u_{p,q} \,, \\
T'_{zx} &= \rho_0(k_1 M_s^3 + k_3 M_s)m_x + c_{zxpq}u_{p,q} \,, \\
T'_{zy} &= \rho_0(k_1 M_s^3 + k_3 M_s)m_y + c_{zypq}u_{p,q} \,, \\
T'_{zz} &= c_{zzpq}u_{p,q} \,,
\end{aligned}
\tag{5.62}
$$

$$
\begin{aligned}
R'_x &= \mu_0 h_x - k_1 M_s^2 m_x - k_1 M_s^3 u_{z,x} - k_3 M_s(u_{x,z} + u_{z,x}) + \alpha_1 m_{x,\ell\ell} \,, \\
R'_y &= \mu_0 h_y - k_1 M_s^2 m_y - k_1 M_s^3 u_{z,y} - k_3 M_s(u_{y,z} + u_{z,y}) + \alpha_1 m_{y,\ell\ell} \,,
\end{aligned}
\tag{5.63}
$$

and

$$
\begin{aligned}
Q'_x &= \rho_0 \alpha_1 m_{x,\ell} n_\ell \,, \\
Q'_y &= \rho_0 \alpha_1 m_{y,\ell} n_\ell \,.
\end{aligned}
\tag{5.64}
$$

In (5.62) and (5.63) M_s denotes the value of the saturated magnetization.

The linearized balance equations are for the case of constant saturated magnetization

$$
\tilde{M}_k m_k = 0 \;; \quad \tilde{M}_k \dot{m}_k = 0 \,, \tag{5.65}
$$

$$
\dot{m}_k = \frac{M_s}{\sqrt{\tilde{R}_k \tilde{R}_k}} \Gamma e_{kpq} m_p \tilde{M}_q + \Gamma e_{kpq} \tilde{M}_p R'_q \,, \tag{5.66}
$$

$$T'_{k\ell,\ell} + \mu_0 h_{k,\ell}\tilde{M}_\ell \rho_0 + \rho_0 f'_k = \rho_0 \ddot{u}_k \ , \tag{5.67}$$

where

$$h_k = h_k^{(0)} + h_k^{(1)} \ . \tag{5.68}$$

In (5.68) $h_k^{(0)}$ is the perturbing external small magnetic field intensity, while $h_k^{(1)}$ is the internal field. We often have

$$h_k^{(1)} = -\phi'_{,k} \ , \tag{5.69}$$

if $\underline{J} = 0$ and we neglect $\underline{e}$, the extra electric field intensity. In this case ϕ' satisfies

$$\Delta\phi' = \rho_0 m_{k,k} - \rho_0 \tilde{M}_k u_{p,pk} \ . \tag{5.70}$$

In (5.67) $\rho_0 f'_k$ is given by

$$\rho_0 f'_k = \tilde{T}_{k\ell,\ell} + \tilde{t}_{k\ell,\ell} + \rho_0 f_k^{(mech)} - \rho_0 u_{p,p} f_k^{(mech)} \ . \tag{5.71}$$

6. Magneto-elastic waves in the infinite space and in the half-space

First we consider a homogeneous, anisotropic infinite space, magnetized to saturation. We shall assume that dissipation may be neglected and that the material is non-conductive.

Strictly speaking an infinite space cannot be magnetized, because there is no exterior field to perform the magnetization process. However, we may consider a homogeneous anisotropic sphere of radius R, placed in an external field and let R tend to infinity. As is well known, homogeneous fields and magnetizations may occur in ellipsoids.

In the rigid sphere we decompose the magnetic field

$$\underline{\tilde{H}} = \underline{\tilde{H}}^{(0)} + \underline{\tilde{H}}^{(1)} , \tag{6.1}$$

where $\underline{\tilde{H}}^{(0)}$ is the external field

$$\tilde{H}_z^{(0)} = H \ ; \quad \tilde{H}_x^{(0)} = \tilde{H}_y^{(0)} = 0 , \tag{6.2}$$

with H prescribed. In the crystalline sphere the internal field may be represented by

$$\tilde{H}_k^{(1)} = -\rho D_{k\ell}\tilde{M}_\ell , \tag{6.3}$$

where $D_{k\ell}$ is the demagnetization tensor, which for the isotropic sphere has the form $\frac{1}{3}\,\delta_{k\ell}$. A crystal of cubic symmetry has the same demagnetization tensor.

As we have

$$\dot{\tilde{M}}_k = \Gamma e_{kpq}\tilde{M}_p\tilde{R}_q = 0 , \tag{6.4}$$

we find for $\underline{\tilde{R}}$

$$\underline{\tilde{M}} = \lambda\underline{\tilde{R}} . \tag{6.5}$$

The factor λ is determined by the condition

$$\tilde{M}_k\tilde{M}_k = M_s^2 . \tag{6.6}$$

With (5.19), (5.46), (5.47) and (5.49) we arrive at the system

$$\tilde{M}_x = \lambda\{-\frac{1}{3}\rho_0\tilde{M}_x - \frac{1}{3}k_1\tilde{M}_x(\tilde{M}_y^2 + \tilde{M}_z^2)\} ,$$

$$\tilde{M}_y = \lambda\{-\frac{1}{3}\rho_0\tilde{M}_y - \frac{1}{3}k_1\tilde{M}_y(\tilde{M}_z^2 + \tilde{M}_x^2)\} , \tag{6.7}$$

$$\tilde{M}_z = \lambda\{\mu_0 H - \frac{1}{3}\rho_0\tilde{M}_z - \frac{1}{3}k_1\tilde{M}_z(\tilde{M}_x^2 + \tilde{M}_y^2)\} ,$$

for a crystalline material with cubic symmetry, if the coordinate axes are taken along the edges of the cube. The system (6.7) has several solutions. One solution is

$$\tilde{M}_x = \tilde{M}_y = 0, \; \tilde{M}_z = M_s , \tag{6.8}$$

while

$$\lambda = \frac{M_s}{\mu_0 H - \frac{1}{3}\rho_0 M_s} . \tag{6.9}$$

We shall take (6.8) with (6.9) as the basic solution, which will be perturbed.

If there is an extra external field $\underline{h}^{(0)}$ and a deformation field, the first order equations for $\underline{m}$ and $\underline{u}$ are (cf. (5.65) to (5.67))

$$\tilde{M}_k\dot{m}_k = 0 , \tag{6.10}$$

$$\dot{m}_k = \frac{\Gamma}{\lambda} e_{kpq} m_p \tilde{M}_q + \Gamma e_{kpq}\tilde{M}_p R'_q , \tag{6.11}$$

$$T'_{k\ell,\ell} + \mu_0 h_{k,\ell}\rho_0\tilde{M}_\ell = \rho_0\ddot{u}_k , \tag{6.12}$$

where

$$\underline{h} = \underline{h}^{(0)} + \underline{h}^{(1)} . \tag{6.13}$$

Written out these equations become

$$\dot{m}_x = \frac{\Gamma}{\lambda} m_y M_s - \Gamma M_s\{\mu_0 h_y^{(0)} - \mu_0\phi'_{,y} - k_1 M_s^2 m_y -$$

$$- k_1 M_s^3 u_{z,y} - k_3 M_s(u_{y,z} + u_{z,y}) + \alpha_1 m_{y,\ell\ell}\} ,$$

$$\dot{m}_y = -\frac{\Gamma}{\lambda} m_x M_s + \Gamma M_s\{\mu_0 h_x^{(0)} - \mu_0\phi'_{,x} - k_2 M_s^2 m_x -$$

$$- k_1 M_s^3 u_{z,x} - k_3 M_s (u_{x,z} + u_{z,x}) + \alpha_1 m_{x,\ell\ell}\} ,$$

$$\phi'_{,kk} = \rho_0(m_{x,x} + m_{y,y}) - \rho_0 M_s u_{\ell,\ell z} , \qquad (6.14)$$

$$\ddot{u}_x = \frac{c_{x\ell pq}}{\rho_0} u_{p,q\ell} + k_3 M_s m_{x,z} - \mu_0\phi'_{,xz} M_s + \mu_0 h^{(0)}_{z,x} M_s ,$$

$$\ddot{u}_y = \frac{c_{y\ell pq}}{\rho_0} u_{p,q\ell} + k_3 M_s m_{y,z} - \mu_0\phi'_{,yz} M_s + \mu_0 h^{(0)}_{z,y} M_s ,$$

$$\ddot{u}_z = \frac{c_{z\ell pq}}{\rho_0} u_{p,q\ell} + (k_1 M_s^3 + k_3 M_s)(m_{x,x} + m_{y,y}) -$$

$$- \mu_0\phi'_{,zz} M_s + \mu_0 h^{(0)}_{z,z} M_s .$$

From (5.56) we derive, with the aid of (5.49) to (5.54) the following values for the constants of elasticity

$$\frac{c_{1111}}{\rho_0} = \frac{c_{2222}}{\rho_0} = c_1 , \qquad (6.15)$$

$$\frac{c_{3333}}{\rho_0} = c_1 + \frac{5}{2} k_2 M_s^2 , \qquad (6.16)$$

$$\frac{c_{1122}}{\rho_0} = \frac{c_{1133}}{\rho_0} = \frac{c_{2233}}{\rho_0} = \frac{c_{2211}}{\rho_0} = c_2 , \qquad (6.17)$$

$$\frac{c_{3311}}{\rho_0} = \frac{c_{3322}}{\rho_0} = c_2 - \tfrac{1}{2} k_2 M_s^2 , \qquad (6.18)$$

$$\frac{c_{1212}}{\rho_0} = \frac{c_{1221}}{\rho_0} = \frac{c_{2121}}{\rho_0} = \frac{c_{2112}}{\rho_0} = c_3 , \qquad (6.19)$$

$$\frac{c_{1313}}{\rho_0} = \frac{c_{2323}}{\rho_0} = c_3 + \tfrac{1}{2} k_2 M_s^2 , \qquad (6.20)$$

$$\frac{c_{1331}}{\rho_0} = \frac{c_{2332}}{\rho_0} = c_3 + k_3 M_s^2 , \qquad (6.21)$$

$$\frac{c_{3131}}{\rho_0} = \frac{c_{3232}}{\rho_0} = c_3 + k_1 M_s^4 + 2k_3 M_s^2 \, , \tag{6.22}$$

$$\frac{c_{3113}}{\rho_0} = \frac{c_{3223}}{\rho_0} = c_3 + \tfrac{1}{2}k_2 M_s^2 + k_3 M_s^2 \, . \tag{6.23}$$

The other $c_{k\ell pq}$ are equal to zero.

The symmetry of the coefficients $c_{k\ell pq}$ shows that the system (6.14) breaks up into two systems if we take $\underline{h}^{(0)} = 0$ and consider waves that propagate along the z-axis. In that case we have $\frac{\partial}{\partial x} = \frac{\partial}{\partial y} = 0$ and (6.14) becomes

$$\begin{aligned}
\dot{m}_x &= \frac{\Gamma M_s}{\lambda} m_y - \Gamma M_s \{-k_1 M_s^2 m_y - k_3 M_s u_{y,z} + \alpha_1 m_{y,zz}\} \, , \\
\dot{m}_y &= -\frac{\Gamma M_s}{\lambda} m_x + \Gamma M_s \{-k_1 M_s^2 m_x - k_3 M_s u_{x,z} + \alpha_1 m_{x,zz}\} \, , \\
\ddot{u}_x &= (c_3 + \tfrac{1}{2}k_2 M_s^2) u_{x,zz} + k_3 M_s m_{x,z} \, , \\
\ddot{u}_y &= (c_3 + \tfrac{1}{2}k_2 M_s^2) u_{y,zz} + k_3 M_s m_{y,z}
\end{aligned} \tag{6.24}$$

and

$$\begin{aligned}
\phi'_{,zz} &= -\rho_0 M_s u_{z,zz} \, , \\
\ddot{u}_z &= (c_1 + \frac{5}{2} k_2 M_s^2) u_{z,zz} - \mu_0 \phi'_{,zz} M_s \, .
\end{aligned} \tag{6.25}$$

Eliminating $\phi'_{,zz}$ from (6.25) we obtain

$$\ddot{u}_z = (c_1 + \frac{5}{2} k_2 M_s^2 + \mu_0 \rho_0 M_s^2) u_{z,zz} \, . \tag{6.26}$$

This is a simple wave equation for the longitudinal wave. The wave velocity is

$$c_\ell = \sqrt{c_1 + \frac{5}{2} k_2 M_s^2 + \mu_0 \rho_0 M_s^2} \, . \tag{6.27}$$

A convenient way to solve the equations (6.24) may be obtained after introducing the quantities

$$m^{\pm} = m_x \pm im_y \; ; \quad u^{\pm} = u_x \pm iu_y \; . \tag{6.28}$$

The system (6.24) now breaks up into two uncoupled systems of equations

$$\begin{aligned} \dot{m}^{+} &= -i \frac{\Gamma M_s}{\lambda} m^{+} - i\Gamma M_s \{k_1 M_s^2 m^{+} + k_3 M_s u^{+}_{,z} - \alpha_1 m^{+}_{,zz}\} \, , \\ \ddot{u}^{+} &= (c_3 + \tfrac{1}{2}k_2 M_s^2) u^{+}_{,zz} + k_3 M_s m^{+}_{,z} \, , \end{aligned} \tag{6.29}$$

$$\begin{aligned} \dot{m}^{-} &= i \frac{\Gamma M_s}{\lambda} m^{-} + i\Gamma M_s \{k_1 M_s^2 m^{-} + k_3 M_s u^{-}_{,z} - \alpha_1 m^{-}_{,zz}\} \, , \\ \ddot{u}^{-} &= (c_3 + \tfrac{1}{2}k_2 M_s^2) u^{-}_{,zz} + k_3 M_s m^{-}_{,z} \, . \end{aligned} \tag{6.30}$$

The frequency equation of (6.29) is

$$\left[\frac{\omega}{\Gamma M_s} - \left(\frac{1}{\lambda} + k_1 M_s^2 + \alpha_1 K_z^2\right)\right]\left[\frac{\omega^2}{K_z^2} - \left(c_3 + \tfrac{1}{2}k_1 M_s^2\right)\right] - k_3^2 M_s^2 = 0 \, , \tag{6.31}$$

while the corresponding equation for (6.30) is

$$\left[\frac{\omega}{\Gamma M_s} + \left(\frac{1}{\lambda} + k_1 M_s^2 + \alpha_1 K_z^2\right)\right]\left[\frac{\omega^2}{K_z^2} - \left(c_3 + \tfrac{1}{2}k_1 M_s^2\right)\right] + k_3^2 M_s^2 = 0 \, . \tag{6.32}$$

In (6.31) and (6.32) ω and K_z denote the frequency and the wave number, respectively, of the harmonic waves along the z-axis.

We are now considering the most elementary boundary value problem, that of the half-space, that we shall assume to occupy the region

$$z \geq 0 \, . \tag{6.33}$$

Again we limit the discussion to the perturbation of the basic solution

$$\tilde{M}_x = \tilde{M}_y = 0, \; \tilde{M}_z = M_s \, , \tag{6.34}$$

that exists if we take one of the edges of the cube along the z-axis.

For an oblique incident harmonic wave that is reflected at the plane $z = 0$, the governing equations are, if we take the plane of incidence to be the x-z-plane

$$\dot{m}_x = \frac{\Gamma M_s}{\lambda} m_y - \Gamma M_s\{-k_1 M_s^2 m_y - k_3 M_s u_{y,z} + \alpha_1 (m_{y,zz} + m_{y,xx})\} ,$$

$$\dot{m}_y = -\frac{\Gamma M_s}{\lambda} m_x + \Gamma M_s\{-\mu_0 \phi'_{,x} - k_1 M_s^2 m_x - k_1 M_s^2 u_{z,x} - k_3 M_s (u_{x,z} + u_{z,x}) + \alpha_1 (m_{x,xx} + m_{x,zz})\} ,$$

$$\phi'_{,kk} = \rho_0 m_{x,x} - \rho_0 M_s (u_{x,xz} + u_{z,zz}) ,$$

$$\ddot{u}_x = \frac{c_{x\ell pq}}{\rho_0} u_{p,q} + k_3 M_s m_{x,z} - \mu_0 \phi'_{,xz} M_s , \tag{6.35}$$

$$\ddot{u}_y = \frac{c_{y\ell pq}}{\rho_0} u_{p,q} + k_3 M_s m_{y,z} ,$$

$$\ddot{u}_z = \frac{c_{z\ell pq}}{\rho_0} u_{p,q} + (k_1 M_s^3 + k_3 M_s) m_{x,x} - \mu_0 \phi'_{,zz} M_s .$$

The boundary conditions are

$$\frac{\partial m_x}{\partial z} = \frac{\partial m_y}{\partial z} = 0, \text{ at } z = 0 , \tag{6.36}$$

$$k_3 M_s m_x + c_{13pq} u_{p,q} = 0 ,$$

$$k_3 M_s m_y + c_{23pq} u_{p,q} = 0 , \tag{6.37}$$

$$c_{33pq}\, u_{p,q} = -\mu_0 \rho_0^2 M_s^2 (u_{x,x} + u_{z,z}) , \text{ at } z = 0 .$$

We look for solutions of the type

$$(m_x, m_y, \phi', u_x, u_y, u_z) = (\bar{m}_x, \bar{m}_y, \bar{\phi}', \bar{u}_x, \bar{u}_y, \bar{u}_z) e^{-i\omega t \pm iK_z z + iK_x x} \tag{6.38}$$

For given ω and K_x, K_z is found from the equation

$$\begin{vmatrix} \frac{i\omega}{\Gamma M_s} & \frac{1}{\lambda}+k_1M_s^2+\alpha_1(K_z^2+K_x^2) & 0 & 0 & \pm ik_3M_sK_z & 0 \\ -\frac{1}{\lambda}-k_1M_s^2-\alpha_1(K_z^2+K_x^2) & \frac{i\omega}{\Gamma M_s} & -\mu_0 iK_x & \pm ik_3M_sK_z & 0 & -iK_x(k_1M_s^3+k_3M_s) \\ i\rho_0K_x & 0 & (K_x^2+K_z^2) & \pm\rho_0K_xK_z & 0 & \rho_0M_sK_z^2 \\ \pm ik_3M_sK_z & 0 & \pm\mu_0M_sK_xK_z & \omega^2-\frac{c_{1111}}{\rho_0}K_x^2-\frac{c_{1313}}{\rho_0}K_z^2 & 0 & \mp\frac{c_{1133}}{\rho_0}K_xK_z\mp\frac{c_{1331}}{\rho_0}K_xK_z \\ 0 & \pm ik_3M_sK_z & 0 & 0 & \omega^2-\frac{c_{2121}}{\rho_0}K_x^2-\frac{c_{2323}}{\rho_0}K_z^2 & 0 \\ iK_x(k_1M_s^2+k_3M_s) & 0 & \mu_0K_z^2M_s & \mp\frac{c_{3311}}{\rho_0}K_xK_z\mp\frac{c_{3113}}{\rho_0}K_xK_z & 0 & \omega^2-K_z^2\frac{c_{3333}}{\rho_0}-K_x^2\frac{c_{3131}}{\rho_0} \end{vmatrix} = 0 \,. \quad (6.39)$$

The equation (6.39) has real coefficients, is even in K_x and K_z and of the sixth order in K_x^2 and K_z^2. For fixed ω and K_x we thus find twelve values for K_z:

$$K_z = \pm K_1, \pm K_2, \pm K_3, \pm K_4, \pm K_5, \pm K_6 . \qquad (6.40)$$

We assume that for real K_x one or more values of K_z are also real. The physical meaning of this assumption is that among (6.38) are some progressive waves. The incident progressive wave is represented by

$$(m_x, m_y, \phi', u_x, u_y, u_z) = (M_x^{(inc)}, M_y^{(inc)}, \phi^{(inc)}, u_x^{(inc)}, u_y^{(inc)}, u_z^{(inc)}) . \; e^{-i\omega t - iK_1 z + iK_x x} , \quad (K_1 \text{ real}) . \qquad (6.41)$$

The expression (6.41) satisfies (6.35). Therefore we may treat $M_x^{(inc)}$ as independent, while the other amplitudes depend on its value.

The reflected waves are represented by

$$(m_x, m_y, \phi', u_x, u_y, u_z) = (M_x^{(\ell)}, M_y^{(\ell)}, \phi^{(\ell)}, u_x^{(\ell)}, u_y^{(\ell)}, u_z^{(\ell)}) . \; e^{-i\omega t + iK_\ell z + iK_x x}$$

$$(\ell = 1,2,\dots,6, \; \mathrm{Im}\, K_\ell \overset{!}{>} 0) . \qquad (6.42)$$

They also obey (6.35) and again the only independent amplitudes may be taken to be

$$M_x^{(\ell)}, \; (\ell = 1,2,\dots,6) .$$

For the determination of these six quantities we have the five boundary conditions (6.36) and (6.37), together with the boundary condition for ϕ'. The first becomes

$$-iK_1 M_x^{(inc)} + \sum_{\ell=1}^{6} iK_\ell M_x^{(\ell)} = 0 , \qquad (6.43)$$

while the other may also be expressed linearly in $M_x^{(\ell)}$ and $M_x^{(inc)}$. We note that the six equations are non-homogeneous. Thus, in general, we may expect these to have one solution.

We now derive the boundary conditions for ϕ'. In $z < 0$ it satisfies

$$\phi'_{,xx} + \phi'_{,yy} = 0 \,, \tag{6.44}$$

with the solution

$$\phi' = Ae^{K_x z} e^{i(K_x x - \omega t)} \,, \quad z < 0 \,. \tag{6.45}$$

By the requirement of continuity at $z = 0$, we find for A

$$A = \phi^{(inc)} + \sum_{\ell=1}^{6} \phi^{(\ell)} \,. \tag{6.46}$$

Since

$$\left(\frac{\partial\phi'}{\partial z}\right)^+ = \left(\frac{\partial\phi'}{\partial z}\right)^- \,, \tag{6.47}$$

we derive the sixth equation in the form

$$(K_x + iK_1)\phi^{(inc)} + \sum_{\ell=1}^{6} (K_x - iK_\ell)\phi^{(\ell)} = 0 \,. \tag{6.48}$$

The solution of the problem (6.35) to (6.37) in the manner indicated is quite involved. Therefore, we can better split the problem into a magnetic one and an elastic one and take into account the coupling afterwards. For more details about this method, we refer to [9].

Acknowledgement:
Part of the work described in this paper formed part of a program supported by a Grant from the National Science Foundation, at Lehigh University, Bethlehem, Pennsylvania.

References

[1] TOUPIN, R.A., J. Rat. Mech. Anal. 5, 849-916 (1956).

[2] TOUPIN, R.A., Int. J. Eng. Sc. 1, 101-126 (1963).

[3] TIERSTEN, H.F., J. Math. Phys. 5, 1298-1318 (1964).

[4] TIERSTEN, H.F., J. Math. Phys. 6, 779-787 (1965).

[5] BROWN, W.F., Magneto-elastic Interactions, Springer-Verlag, Berlin, 1966.

[6] AKHIESER, A.T., BAR'YAKHTAR, V.G., PELETMINSKII, S.V., Spin waves, North Holl. Publ. Co., Amsterdam, 1968.

[7] ALBLAS, J.B., Mechanics of Generalized Continua, 350-354, Springer-Verlag, 1967.

[8] ALBLAS, J.B., Symp. Math. I, 229-251, Inst. Naz. di alta Math. ed. Ac. Press (1968).

[9] ALBLAS, J.B., Dynamic Theory of Magneto-elastic Interactions, 1-159, unpublished lecture notes.

[10] FANO, R.M., CHU, L.J., and ADLER, R.B., Electro-magnetic fields, Energy and Forces, Wiley & Sons, New York, 1960.

[11] PENFIELD, P., HAUS, H.A., Electrodynamics of moving media, M.I.T. press, Cambridge, Mass., 1967.

[12] GREEN, A.E., RIVLIN, R.S., Arch. Rat. Mech. Anal., 16, 325-353, 1964.

PLASTICITY AND CREEP THEORY IN ENGINEERING MECHANICS

J.F. BESSELING, Delft

1. Introduction

In October 1675 Robert Hooke showed in London an "Experiment of Spring" to the King, who was "very well pleased" /1/. Robert Hooke was Curator of Experiments of the Royal Society. The business of the Curator was "to furnish the Society every day they meet - that was once a week - with three or four considerable experiments, expecting no recompence till the Society gett a stock enabling to give it". Hooke was quite intimate with the King, who took a detailed interest in scientific proceedings. Much of the work bore upon the needs of navigators at sea and it was in connection with the use of springs in time-keepers that Robert Hooke enunciated the fundamental law of spring.

In 1833 Vicat in France published results of creep experiments on iron wires /2/. He was well aware of the large influence of the temperature on the creep-phenomenon, since he recorded the temperature in the laboratory within 0.1 ^{o}K.

Both the springaction as well as the ductility have been known properties of metals from the day on they were being used. Quantitative data on their mechanical behaviour, for which a description with the precision of mathematics has to be sought, are still not abundant however. In fact in mechanics, with regard to the mechanical behaviour of materials, theory has flourished and experiment has withered.

The theorists have provided us with innumerable concepts and models of material behaviour. Few of them were prompted by direct observation of experimental data and most of them were given a place in the literature before an experimental verification of their descriptive value was even attempted.

It also could happen that even in otherwise outstanding text-books on the mechanics of solids there appears an illustration of the Bauschinger-effect, showing an highly anomalous stress-strain diagram, of which I am sure that it is not related to a real test.

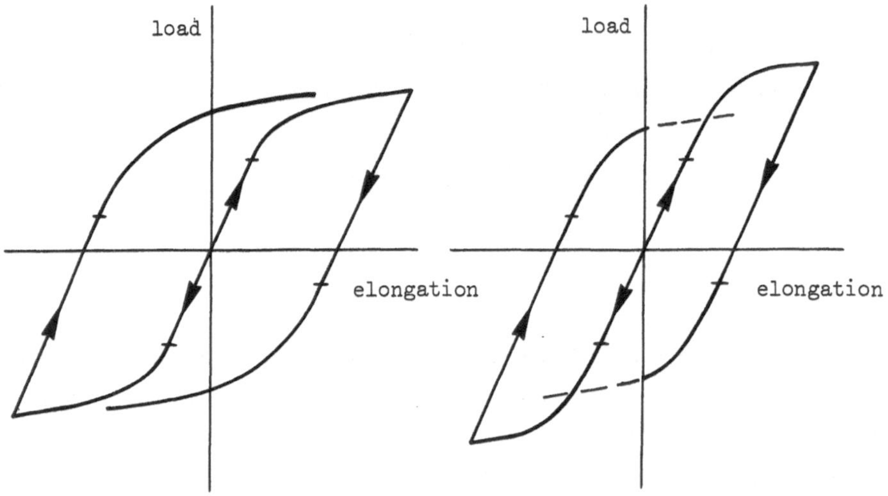

Fig. 1. Normal behaviour. Textbook illustration.

I think in the mechanics of solids we have been spoilt by the far reaching and accurate deductions that could be drawn from the theory of elasticity on the basis of the knowledge of a few physical constants. This success was due to the fact that the theory of elasticity deals with a reversible sequence of states. In the more general deformation problem the theoretical and experimental research has to cope with a sequence of states in an irreversible process. Not only the state of motion, but the very properties of the material are dependent on the history of the external action.

Plasticity and creep-theory deserves a place in engineering mechanics if it meets the basic requirement: the theory should on the basis of uniquely defined information encompass features of material behaviour that are important in design problems. The proof of strength and serviceability, as it is called for in engineering, rests upon a comparison of numbers, obtainable and obtained in an objective, unequivocal manner. Any concept or model of material behaviour without a recipe, how to determine the governing constants or functionals, is of no use in engineering mechanics.

In my opinion at the moment the model of material behaviour, that can be confidently related to experimental data for a particular material, is the model of so-called ideal creep and ideal plasticity. This model gives a description that is by far not complete, but it is adequate for the calculation of certain limiting states, essential in some design problems.

We obtain more sophisticated models of material behaviour, which still can be related in an unique fashion to experimental data for a particular material, by considering in an element of volume parallel arrangements of a small number of subelements, each of which shows ideal creep or ideal plasticity. Once their accuracy for a quantitative description of the stress-strain relations for a material has been proved by careful experimentation, they may show their value in problems, where the actual stress-distribution at any moment is of prime importance, such as in the problem of safety against crack propagation. Further in terms of uniquely defined models of material behaviour strength requirements may be formulated which provide a more realistic basis for the assessment of the acceptability of a design than requirements with respect to the elastic stress-distribution only.

2. Limit Analysis

At temperatures far below the melting temperature the response of metals is useally nearly time-independent. There is a limited range of stress, in which stress and strain are related by Hooke's law. The elastic moduli are little affected by the history of deformation. The elastic range does vary, however, because of the so-called strain-hardening and the accompanying anisotropic Bauschinger-effect. These are phenomena encountered in any conglomerate of elements that have different resistances against plastic deformation under a certain type of loading. Both on the microscopic scale in the polycristalline material as well as on the macroscopic scale in the various structural elements inhomogeneous plastic deformation leads to internal stresses, that produce a displacement of the elastic range with respect to the unloaded state.

Fig. 2 gives an illustration of strainhardening and Bauschinger-effect for the case that in the material only one component of stress with the associated component of strain, and that in the structural element only one loading parameter with the associated deformation para-

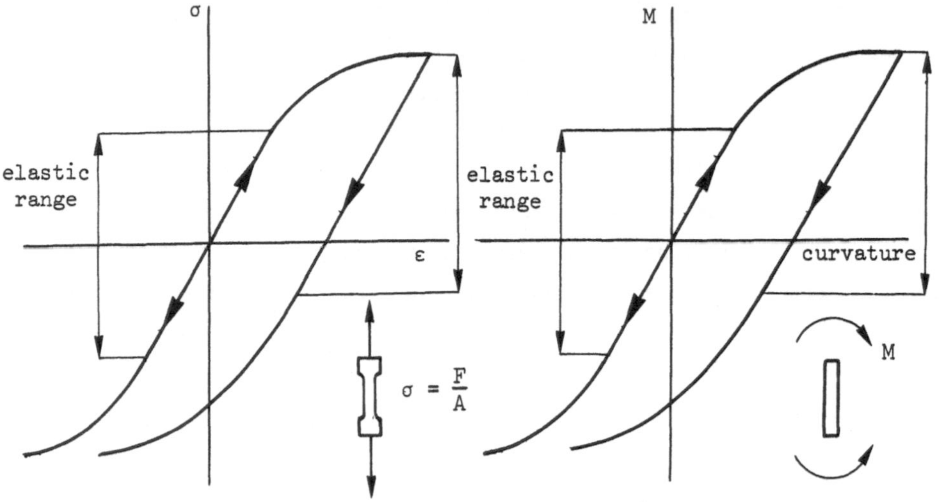

Fig. 2. Strainhardening and Bauschinger-effect.

meter are to be considered. We cannot picture multiaxial states of loading, but we could look upon Fig. 2 as graphs for suitably defined representative loading and deformation parameters, both for the material and for the structural elements.

The analysis of the loading of a structure may be carried out at various levels of approximation. In the continuum approach the material behaviour is often replaced by the stress-strain relations of the elastic-ideally plastic model (Fig. 3 - line a). This leads to an overestimation of the elastic limit for the structural elements, but it does not influence their ultimate load carrying capacity but for the choice of σ^y.

As the next level of approximation we may take finite structural elements to be elastic-ideally plastic with respect to their loading and deformation parameters (Fig. 3 - line b). By taking larger and larger structural elements we get a further overestimation of the elastic limit and of the elastic range of the structure, but the ultimate load carrying capacity is not affected except by the yield limits of the individual finite elements.

The elastic-ideally plastic model eliminates the history of the deformation as a determining factor, both when applied at the material level as well as when applied at the structural level. This simplifies the analysis enormously, while still the for design purposes most important

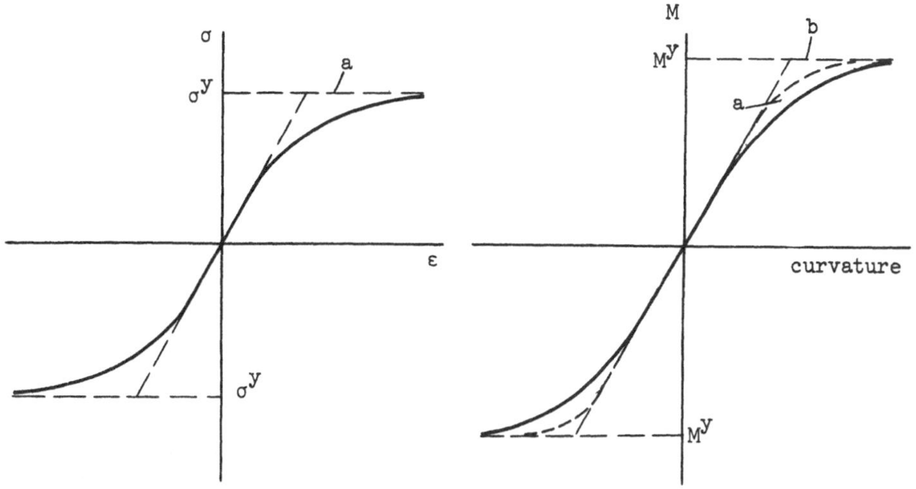

Fig. 3. Two levels of approximation (a and b).

features of ductile material are retained: a finite elastic range of stress with the possibility of large plastic deformation if the stress reaches the yield limit.

If the yield limits of the elements in a structure can be represented by linear conditions for their loading parameters, then the theorems of limit analysis find their mathematical expression in terms of the following dual problems of linear programming algebra /3,4,5,6/.

I Maximize $h^T x$

under the conditions $Ax \leq g$, $x \geq 0$, (2.1.a)

II Minimize $g^T y$

under the conditions $A^T y \geq h$, $y \geq 0$. (2.1.b)

Here h and g are known vectors and A is a known matrix.

We shall illustrate this formulation for the plane framework of Fig. 4. If they have a rectangular cross section the yield limit of each of the four beam elements is determined by

$$\left(\frac{N}{N^y}\right)^2 + \frac{|M_p|}{M_p^y} = 1, \quad \left(\frac{N}{N^y}\right)^2 + \frac{|M_q|}{M_q^y} = 1. \tag{2.2}$$

Since in the structure under consideration the normal force N will remain well below the value of the fully plastic load N^y we may approximate the load carrying capacity of the elements by linear inequalities.

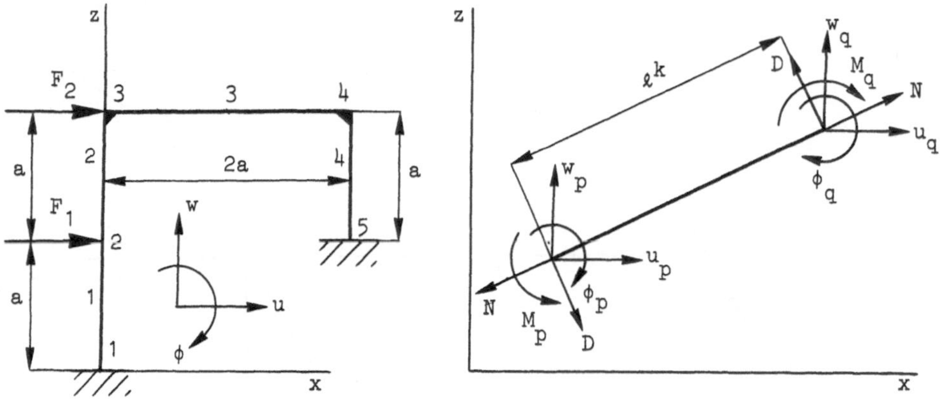

Fig. 4. Plane framework and finite element between nodal points.

$$|M_p| \leq M_p^y \ , \quad |M_q| \leq M_q^y \ . \tag{2.3}$$

Here M^y is the value of the so-called fully plastic moment at which a plastic hinge will develop.

We have defined the state of loading of a beam element in terms of the load parameters

$$(\sigma^k)^T = |N, \ M_p/\ell^k, \ M_q/\ell^k|. \tag{2.4}$$

The virtual work of deformation is then given by the scalar product of the vectors σ^k and $\delta\varepsilon^k$, where

$$\varepsilon^k = D^k u^k. \tag{2.5}$$

The vector ε^k represents the vector of three deformation parameters, defined in terms of the six nodal displacements and rotations of the element, that form together the vector u^k.

For an external nodal load vector f equilibrium of the structure requires that the equality

$$\sigma^T \delta\varepsilon = \sigma^T D\delta u = f^T \delta u \tag{2.6}$$

holds for all kinematically admissible vectors of virtual nodal displacements and rotations, δu. Hence the vector of <u>all</u> load parameters in the structure, σ, must satisfy the equilibrium equations

$$D^T \sigma = f, \tag{2.7}$$

while continuity of the structure will be maintained, provided <u>all</u> deformation parameters, ε, are related to the nodal displacements and rotations by

$$Du = \varepsilon. \tag{2.8}$$

The matrix D and the vectors u, f, σ and ε for the structure of Fig. 4 are given in table 1.

$$u^T = | u_2 \quad w_2 \quad \phi_2 a \quad u_3 \quad w_3 \quad \phi_3 a \quad u_4 \quad w_4 \quad \phi_4 a |,$$

$$\varepsilon = \begin{vmatrix} \varepsilon_1^1 \\ \varepsilon_2^1 \\ \varepsilon_3^1 \\ \varepsilon_1^2 \\ \varepsilon_2^2 \\ \varepsilon_3^2 \\ \varepsilon_1^3 \\ \varepsilon_2^3 \\ \varepsilon_3^3 \\ \varepsilon_1^4 \\ \varepsilon_2^4 \\ \varepsilon_3^4 \end{vmatrix}, \quad D = \begin{bmatrix} 0 & 1 & 0 & 0 & 0 & 0 & 0 & 0 & 0 \\ 1 & 0 & 0 & 0 & 0 & 0 & 0 & 0 & 0 \\ -1 & 0 & 1 & 0 & 0 & 0 & 0 & 0 & 0 \\ 0 & -1 & 0 & 0 & 1 & 0 & 0 & 0 & 0 \\ -1 & 0 & -1 & 1 & 0 & 0 & 0 & 0 & 0 \\ 1 & 0 & 0 & -1 & 0 & 1 & 0 & 0 & 0 \\ 0 & 0 & 0 & -1 & 0 & 0 & 1 & 0 & 0 \\ 0 & 0 & 0 & 0 & 1 & -2 & 0 & -1 & 0 \\ 0 & 0 & 0 & 0 & -1 & 0 & 0 & 1 & 2 \\ 0 & 0 & 0 & 0 & 0 & 0 & 0 & 1 & 0 \\ 0 & 0 & 0 & 0 & 0 & 0 & 1 & 0 & -1 \\ 0 & 0 & 0 & 0 & 0 & 0 & -1 & 0 & 0 \end{bmatrix}, \quad \sigma = \begin{vmatrix} N^1 \\ M_1^1/a \\ M_2^1/a \\ N^2 \\ M_2^2/a \\ M_3^2/a \\ N^3 \\ M_3^3/(2a) \\ M_4^3/(2a) \\ N^4 \\ M_4^4/a \\ M_5^4/a \end{vmatrix},$$

$$f^T = | F_1 \quad 0 \quad 0 \quad F_2 \quad 0 \quad 0 \quad 0 \quad 0 \quad 0 |.$$

Table 1. Structural matrix and vectors.

Since in the various problems of limit analysis only the bending moments are subject to the inequalities (2.3) we may eliminate from (2.7) and (2.8) respectively the normal forces and the associated elongations of the beam elements. The resulting system of vectors and matrix is given in table 2.

$$u^T = | \; u_2 \quad \phi_2 a \quad \phi_3 a \quad u_4 \quad \phi_4 a \; |,$$

$$\varepsilon = \begin{vmatrix} \varepsilon_2^1 \\ \varepsilon_3^1 \\ \varepsilon_2^2 \\ \varepsilon_3^2 \\ \varepsilon_2^3 \\ \varepsilon_3^3 \\ \varepsilon_2^4 \\ \varepsilon_3^4 \end{vmatrix}, \quad D = \begin{bmatrix} 1 & 0 & 0 & 0 & 0 \\ -1 & 1 & 0 & 0 & 0 \\ -1 & -1 & 0 & 1 & 0 \\ 1 & 0 & 1 & -1 & 0 \\ 0 & 0 & -2 & 0 & 0 \\ 0 & 0 & 0 & 0 & 2 \\ 0 & 0 & 0 & 1 & -1 \\ 0 & 0 & 0 & -1 & 0 \end{bmatrix}, \quad \sigma = \begin{vmatrix} M_1^1/a \\ M_2^1/a \\ M_2^2/a \\ M_3^2/a \\ M_3^3/(2a) \\ M_4^3/(2a) \\ M_4^4/a \\ M_5^4/a \end{vmatrix},$$

$$f^T = | \; F_1 \quad 0 \quad 0 \quad F_2 \quad 0 \; |.$$

Table 2. Reduced structural matrix and vectors.

By the first collapse theorem collapse of the structure at a load vector λf will not take place below the maximum value of λ, that satisfies the conditions

$$D^T\sigma - \lambda f = 0,$$

$$\sigma_i^+ \le M_i^y/\ell^k = \sigma_i^y, \quad \sigma_i^- \le M_i^y/\ell^k = \sigma_i^y, \tag{2.9}$$

where

$$\sigma_i^+ = \sigma_i \text{ if } \sigma_i \ge 0, \text{ else } \sigma_i^+ = 0,$$

$$\sigma_i^- = -\sigma_i \text{ if } \sigma_i \le 0, \text{ else } \sigma_i^- = 0. \tag{2.10}$$

According to the second collapse theorem collapse will take place for all values of λ above the minimum value of λ determined by

$$\lambda f^T\dot{u} = (\sigma^y)^T \dot{\varepsilon}^+ + (\sigma^y)^T \dot{\varepsilon}^-, \tag{2.11}$$

where $\dot{u}$ and $\dot{\varepsilon}$ satisfy the conditions

$$\dot{\varepsilon} - D\dot{u} = 0, \; f^T\dot{u} \ge 1. \tag{2.12}$$

Here $\dot{\varepsilon}^+$ and $\dot{\varepsilon}^-$ are defined by

$$\dot{\varepsilon}_i^+ = \dot{\varepsilon}_i \quad \text{if} \quad \dot{\varepsilon}_i \geq 0, \quad \text{else} \quad \dot{\varepsilon}_i^+ = 0,$$
$$\dot{\varepsilon}_i^- = \dot{\varepsilon}_i \quad \text{if} \quad \dot{\varepsilon}_i \leq 0, \quad \text{else} \quad \dot{\varepsilon}_i^- = 0. \tag{2.13}$$

The problem of the determination of the collapse load for the structure of Fig. 4 is now contained in the dual problems of linear programming algebra (2.1a) and (2.1b) with the following matrix and vectors

$$x^T = |(\sigma^+)^T \quad (\sigma^-)^T \quad \lambda|,$$

$$y = \begin{vmatrix} \dot{\varepsilon}^+ \\ \dot{\varepsilon}^- \\ \dot{u}^+ \\ \dot{u}^- \end{vmatrix}, \quad A = \begin{bmatrix} I & -I & 0 \\ -I & I & 0 \\ -D^T & D^T & f \\ D^T & -D^T & -f \end{bmatrix}, \quad g = \begin{vmatrix} \sigma^y \\ \sigma^y \\ 0 \\ 0 \end{vmatrix},$$

$$h^T = |\, 0 \quad 0 \quad 1 \,|.$$

Table 3. L.P. matrix and vectors for collapse load.

It should be noted that even for the simple structure of Fig. 4 the matrix A has 26 rows and 17 columns.

In the case of fluctuating loads the shakedown theorems can provide us with the loads for which ultimately a purely elastic response of the structure will be obtained.

Suppose that according to an elastic calculation the load parameters of the finite elements in the structure would fluctuate between values $\lambda\sigma^{max}$ and $\lambda\sigma^{min}$.

The <u>first shakedown theorem</u> now states that for this one parameter type of loading the structure will shake down to time independent plastic strains for values of λ, that remain below the maximum value of λ, satisfying the conditions

$$D^T\sigma^r = 0,$$
$$\lambda\sigma_i^{max} + \sigma_i^r \leq \sigma_i^y, \quad \lambda\sigma_i^{min} + \sigma_i^r \geq -\sigma_i^y. \tag{2.14}$$

The <u>second shakedown theorem</u> complements this lower bound for the shakedown load by the statement: the structure will not shakedown to time independent plastic strains for values of λ above the minimum value of λ, determined by

$$\lambda[(\sigma^{max})^T\Delta\varepsilon^+ - (\sigma^{min})^T\Delta\varepsilon^-] = (\sigma^y)^T\Delta\varepsilon^+ + (\sigma^y)^T\Delta\varepsilon^-, \tag{2.15}$$

where $\Delta\varepsilon$ satisfies the conditions

$$\Delta\varepsilon - D\Delta u = 0, \quad (\sigma^{max})^T \Delta\varepsilon^+ - (\sigma^{min})^T \Delta\varepsilon^- \geq 1. \qquad (2.16)$$

In terms of the dual problems of linear programming algebra (2.1.a) and (2.1.b) the determination of the shakedown load of the structure entails the following matrix and vectors (table 4)

$$x^T = |\sigma^{r+} \quad \sigma^{r-} \quad \lambda \;|,$$

$$y = \begin{vmatrix} \Delta\varepsilon^+ \\ \Delta\varepsilon^- \\ \Delta u^+ \\ \Delta u^- \end{vmatrix}, \quad A = \begin{bmatrix} I & -I & \sigma^{max} \\ -I & I & -\sigma^{min} \\ -D^T & D^T & 0 \\ D^T & -D^T & 0 \end{bmatrix}, \quad g = \begin{vmatrix} \sigma^y \\ \sigma^y \\ 0 \\ 0 \end{vmatrix},$$

$$h^T = |\; 0 \quad 0 \quad 1 \;|.$$

Table 4. L.P. matrix and vectors for shakedown load.

We notice that the matrix A for the structure of Fig. 4 has again 26 rows and 17 columns. It should be realized, however, that for the determination of σ^{max} and σ^{min} a detailed elastic analysis of the structure has to be carried out.

Finally we draw attention to the minimum weight design, based upon the concept of ideal plasticity.

We take the expression

$$M_p^y \ell^k + M_q^y \ell^k = \sigma_p^y (\ell^k)^2 + \sigma_q^y (\ell^k)^2 \qquad (2.17)$$

as a measure for the contribution of a beam element in a plane framework to the weight of the structure. This is justified if the beam elements have a given length and a given form of cross section. We introduce a vector ℓ^2, which contains for each beam element two elements, equal to the square of the length of the beam element under consideration. The minimum weight problem, for instance for the structure of Fig. 4, can then be formulated by:

$$\text{minimize } (\sigma^y)^T \ell^2 \qquad (2.18)$$

under the conditions

$$D^T\sigma = \lambda f, \quad \sigma^+ \leq \sigma^{y+}, \quad \sigma^- \leq \sigma^{y-}. \qquad (2.19)$$

This is problem (2.1.b) of the dual problems (2.1.a) and (2.1.b) of linear programming algebra, defined by the matrix and vectors

$$x^T = | \dot{u}^+ \quad \dot{u}^- \quad \dot{\varepsilon}^+ \quad \dot{\varepsilon}^- |,$$

$$y = \begin{vmatrix} \sigma^+ \\ \sigma^- \\ \sigma^{y+} \\ \sigma^{y-} \end{vmatrix}, \quad A = \begin{bmatrix} D & -D & -I & 0 \\ -D & D & 0 & -I \\ 0 & 0 & I & 0 \\ 0 & 0 & 0 & I \end{bmatrix}, \quad g = \begin{vmatrix} 0 \\ 0 \\ \ell 2 \\ \ell 2 \end{vmatrix},$$

$$h^T = |\lambda f^T \; -\lambda f^T \quad 0 \quad 0 |.$$

Table 5. L.P. matrix and vectors for minimum weight.

We observe that in the solution of problem (2.1.a) the absolute values of the angular velocities at the plastic hinges are determined by the lengths of the corresponding beam elements (Foulkes mechanism /6/).

Even with the most efficient algorithm the solution of problems of limit analysis for a structure, as formulated above in terms of dual problems of linear programming algebra, requires a much larger memory and takes much more computertime than an elastic analysis for the same structure. Being familiar with the mechanical behaviour of structures we usually can indicate a simpler way to arrive at a satisfactory solution of the problems of limit analysis. For instance a judicious estimate of the location of the plastic hinges will lead us to a quick approximation of the collapse load. However none of these other approaches are systematical to the same extent as the linear programming algebra formulation. Computer programs therefore generally have to be based upon the latter approach if the theorems of limit analysis are taken as a starting point. However in particular for the determination of the collapse load of a structure we have as an alternative approach a complete elastic-plastic analysis up to the moment of collapse.

The piecewize linear analysis for the model of material behaviour, contained in the plastic hinge concept, does not pose any grave computational problems. Further for more complicated models with non-linear yield criteria the linear programming algebra approach is no longer applicable, while computer programs for a complete non-linear analysis are feasible and in certain forms already operational.

For the structural optimization problem, of which the minimum weight problem considered above is the simplest example, of course an entirely different development is required, which we shall not consider here any further.

3. Models of material behaviour

In a continuum description as well as in a finite element approach we introduce a vector of stress components or load parameters σ_i and an associated vector of strain components or deformation parameters ε_i, such that $\sigma_i \dot{\varepsilon}_i$ represents the rate of work of deformation per unit volume or per finite element [1].

In the elastic range the quantities σ_i and ε_i are related by a symmetric matrix of elasticity coëfficiënt S_{ij}. If creep and/or plasticity occurs we introduce a vector of stress-free deformations ε_i''. We may then write

$$\sigma_i = S_{ij}\,\varepsilon_j', \quad \varepsilon_j' = \varepsilon_j - \varepsilon_j''. \tag{3.1}$$

The total strains or deformations ε_i are expressed into the components of the displacement field or into the nodal coordinates u_k.

$$\varepsilon_i = D_i(u_k). \tag{3.2}$$

For the stress-free deformations ε_i'', in view of the irreversibility of the process, only rate equations may be expected. If we do not take the history of deformation into account by the introduction of functionals in time, then only the concepts of ideal creep and ideal plasticity can be our starting point. This implies that the state of the material, as determined by stress, temperature and possibly some additional parameters of state, shall be sufficient to define the matrix coefficients Y_{ij} and the vector quantities c_j in

$$\dot{\sigma}_i = (S_{ij} - Y_{ij})(\dot{\varepsilon}_j - c_j). \tag{3.3}$$

Here c_j represents the creep effects. The coefficients Y_{ij} differ from zero only if time independent plastic deformation takes place.

For the creep effects we take as a basic postulate the existence of an energy dissipation function, which is solely a function of the state of the material, and within the concept of ideal creep a function of stress and temperature

$$\sigma_i \dot{\varepsilon}_i'' = f(\sigma_k, T). \tag{3.4}$$

---

[1] Throughout the paragraphs 3 and 4 we shall use index notation and summation convention with respect to all lower indices.

It may happen that not all deformation components ε_i can give rise to stress-free deformations ε_i''. So we know for instance that for metals no stress-free changes of volume are possible. Let either by experiment or by induction be found that

$$\alpha_{ij}\varepsilon_j'' = 0. \tag{3.5}$$

If we now introduce stress quantities

$$s_i = \sigma_i - \sigma_j\alpha_{ji} \ , \tag{3.6}$$

then we see from

$$\sigma_i\dot{\varepsilon}_i'' = s_i\dot{\varepsilon}_i'' = f, \tag{3.7}$$

that $\dot{\varepsilon}_i''$ are components of a vector in s_i-space. We adopt the associative law and write for the case of creep

$$\dot{\varepsilon}_i'' = \mu \frac{\partial f}{\partial s_i} = c_i . \tag{3.8}$$

The factor μ immediately follows from

$$s_i\dot{\varepsilon}_i'' = \mu\, s_i \frac{\partial f}{\partial s_i} = f, \quad \text{or } \mu = \left(s_i \frac{\partial f}{\partial s_i}\right)^{-1} f. \tag{3.9}$$

For the time independent deformations the basic postulate is the existence of a yield surface in stress space, inside of which we have $Y_{ij} = 0$. Let this surface be determined by

$$\phi(s_i) = 0. \tag{3.10}$$

We might envisage this yield surface as the limiting surface, which among the surfaces of constant energy dissipation represents the surface in s_i-space, where the rate of energy dissipation becomes indeterminately large (Fig. 5).

Again by the associative law we write for the combination of creep and time independent plasticity

$$\dot{\varepsilon}_i'' = \mu \frac{\partial f}{\partial s_i} + p \frac{\partial \phi}{\partial s_i} \ , \tag{3.11}$$

where

$$p \geq 0 \quad \text{only if} \quad \phi = 0 \quad \text{and} \quad \dot{\phi} = 0, \tag{3.12}$$

otherwise $p \equiv 0$.

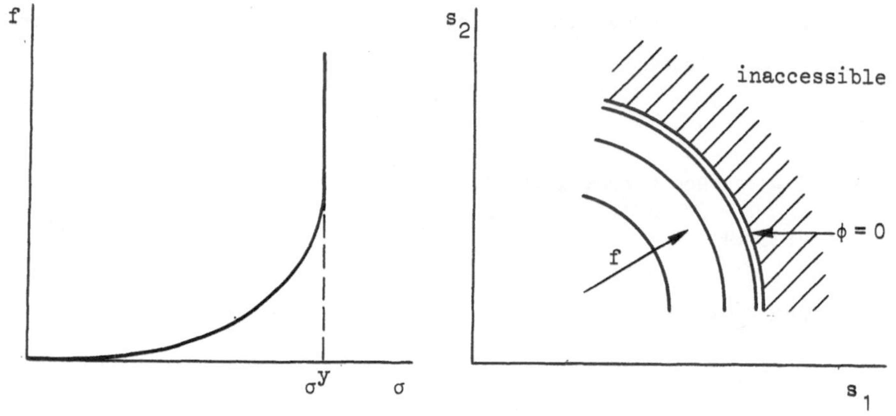

Fig. 5. Energy dissipation function and yield surface.

We determine the factor p from the condition

$$\dot{\phi} = \frac{\partial \phi}{\partial s_i}\,\dot{s}_i = \frac{\partial \phi}{\partial s_i}\,S_{ij}\left(\dot{\varepsilon}_j - \mu\,\frac{\partial f}{\partial s_j} - p\,\frac{\partial \phi}{\partial s_j}\right) = 0. \tag{3.13}$$

With

$$p = \frac{\dfrac{\partial \phi}{\partial s_i}\,S_{ij}\left(\dot{\varepsilon}_j - \mu\,\dfrac{\partial f}{\partial s_j}\right)}{\dfrac{\partial \phi}{\partial s_m}\,S_{mn}\,\dfrac{\partial \phi}{\partial s_n}}, \tag{3.14}$$

we have derived for the case of ideal creep and ideal plasticity the following expressions for the components of the creep vector c_i and for the coefficients of the matrix Y_{ij} in (3.3):

$$c_i = \left(s_m\,\frac{\partial f}{\partial s_m}\right)^{-1} f\,\frac{\partial f}{\partial s_i}, \tag{3.15}$$

$$Y_{ij} = \frac{S_{ik}\,\dfrac{\partial \phi}{\partial s_k}\,\dfrac{\partial \phi}{\partial s_\ell}\,S_{\ell j}}{\dfrac{\partial \phi}{\partial s_m}\,S_{mn}\,\dfrac{\partial \phi}{\partial s_n}}, \tag{3.16}$$

provided

$$\phi(s_i) = 0 \quad \text{and} \quad \frac{\partial \phi}{\partial s_i}\,S_{ij}\,(\dot{\varepsilon}_j - c_j) \geq 0, \tag{3.17}$$

otherwise $Y_{ij} = 0$.

This simple model of ideal creep and ideal plasticity gives a description of material behaviour, which for a uniaxial stress σ is based upon the yield function

$$\phi = \sigma^2 - (\sigma^y)^2, \tag{3.18}$$

while the creep behaviour is then often represented by the power law

$$f = \gamma\left(\frac{\sigma}{\sigma^y}\right)^{2n}. \tag{3.19}$$

According to the relations (3.3) with (3.15), (3.16) and (3.17) we now have the following possibilities:

$$\dot{\sigma} = 0 \text{ provided: } \sigma^2-(\sigma^y)^2 = 0 \text{ and } \sigma\dot{\varepsilon} - \gamma \geq 0, \tag{3.20}$$

otherwise

$$\dot{\sigma} = E\left(\dot{\varepsilon} - \frac{\gamma}{\sigma^y}\left(\frac{\sigma}{\sigma^y}\right)^{2n-1}\right) \tag{3.21}$$

From (3.21) we have in the case of creep at constant stress

$$\dot{\varepsilon} = \frac{\gamma}{\sigma^y}\left(\frac{\sigma}{\sigma^y}\right)^{2n-1}, \tag{3.22}$$

and in the case of relaxation at constant strain

$$\dot{\sigma} = -\frac{E\gamma}{\sigma^y}\left(\frac{\sigma}{\sigma^y}\right)^{2n-1} \tag{3.23}$$

The material constants σ^y, γ and n are to be determined from testdata.

In the continuum description for isotropic materials the yield function is often defined by the Von Mises criterion

$$\phi = 3J_2 - (\sigma^y)^2, \quad J_2 = \tfrac{1}{2}s_{ij}s_{ij}. \tag{3.24}$$

Here s_{ij} are the components of the stress deviator ($s_{ij} = \sigma_{ij} - \frac{1}{3}\sigma_{kk}\delta_{ij}$). The energy dissipation function, leading to the power law (3.19), is then given by

$$f = \gamma\left\{\frac{3J_2}{(\sigma^y)^2}\right\}^n. \tag{3.25}$$

Bauschinger-effect, primary creep and creep recovery are phenomena, which are missing in this description. A simple remedy of this deficiency was already indicated by Wartenberg /7/.

We can envisage a model of material behaviour, in which the rate of work of deformation per unit volume or per finite element is the sum of the contributions of a finite number of subelements.

$$\sigma_i \dot{\varepsilon}_i = \sum_p \sigma_i^p \dot{\varepsilon}_i^p \, . \tag{3.26}$$

Next we establish an unequivocal relationship between the deformation parameters of the subelements, ε_i^p, and the deformation parameters of the conglomerate, ε_i (3.2).

$$\varepsilon_i^p = A_{ik}^p \, \varepsilon_k . \tag{3.27}$$

Now we have

$$\sigma_i \dot{\varepsilon}_i = \left(\sum_p \sigma_i^p A_{ik}^p \right) \dot{\varepsilon}_k$$

or

$$\sigma_i = \sum_p A_{ki}^p \, \sigma_k^p \, . \tag{3.28}$$

For the stress or load parameters of each subelement we have rate equations of the type (3.3)

$$\dot{\sigma}_i^p = (S_{ij}^p - Y_{ij}^p)\,(\dot{\varepsilon}_j^p - c_j^p) \tag{3.29}$$

With $\varepsilon_i^p = \varepsilon_i$, but by giving the various subelements different yield limits and creep constants, we get in the continuum theory a description of the Bauschinger-effect, of primary creep and of creep recovery, as was pointed out in /8/. We can also cope with singular yield surfaces that contain edges or corners. We then consider a number of intersecting yield surfaces. Further it may be observed that with two subelements, one elastic-ideally plastic and one purely elastic, we find a so-called "kinematic hardening" type of model (Fig. 6).

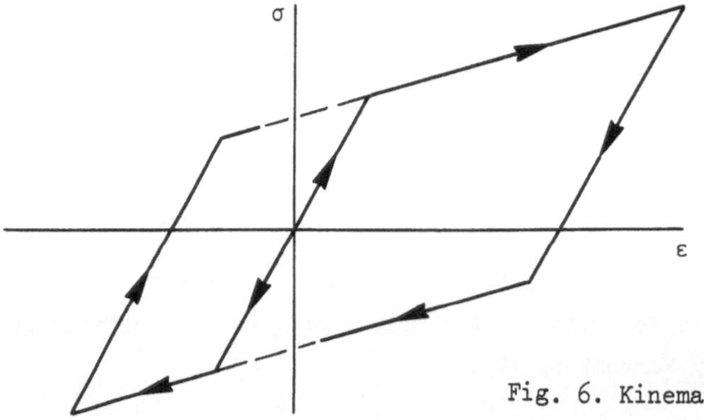

Fig. 6. Kinematic hardening.

In the finite element approach of deformation problems it is useful to consider in each element subdomains, if the finite element that is being applied has a nonuniform stressfield, such as for instance the TRIM-6 plate element (Fig. 7).

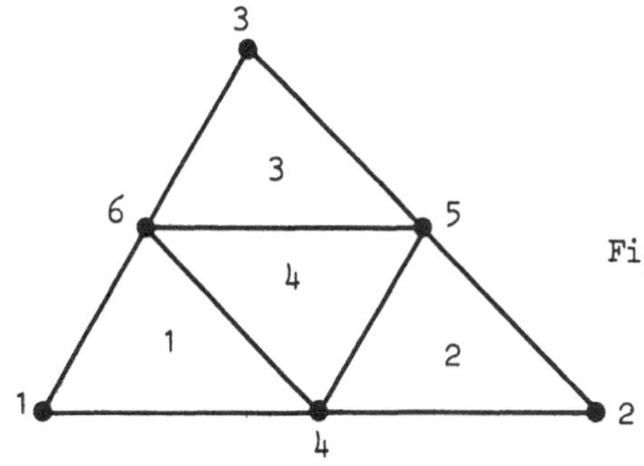

Fig. 7. TRIM-6 plate element.

After a subdivision of the triangular element into four identical triangular domains we establish the relation (3.27). In the elastic-plastic analysis we now find a distribution of plastic strain, which is no longer a linear interpolation over the whole TRIM-6 element, but which consists of linear interpolations within each subdomain separately.

The model based upon (3.26) and (3.27) leads to the development of internal stresses in the course of plastic deformation, that are responsible for strainhardening and Bauschinger-effect as depicted in Fig. 2.

By combining a few elastic-ideally plastic subelements with ideal creep in one subelement we can also accommodate testdata, as recently found in our laboratory at Delft for a steel, that is being used for the construction of pressure vessels of nuclear reactors. This steel shows a definite strainrate dependence in tensile and torsion tests, accompanied by a limited amount of stress relaxation as soon as the deformation is kept constant (Fig. 8).

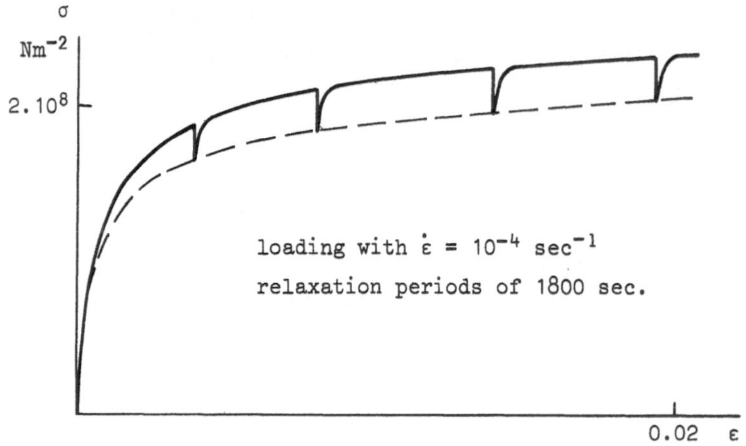

Fig. 8. Time dependent stress-strain data.

4. The geometrically and physically non-linear structural equations

Because of the highly non-linear nature of the creep and plasticity phenomena there are but few analytic solutions of inelastic deformation problems, with the exception of problems in linear visco-elasticity. Therefore computerprograms for numerical solutions have been developed, mostly based on the finite element method.

Let us consider a structure with n nodal coordinates and m generalized strains ε_i of the finite elements in which the structure is divided. Generally we shall have non-linear relations

$$\varepsilon_i = D_i(u_k), \tag{4.1}$$

where now an element ε_i will depend only on those coordinates among u_k, that correspond to the nodal points of the element to which ε_i belongs.

According to the principle of virtual work for external loads f_k, associated with virtual displacements and rotations δu_k, the structure will be in a state of equilibrium only if

$$\sigma_i \delta\varepsilon_i = f_k \delta u_k \tag{4.2}$$

holds for all kinematically admissible δu_k.

We denote the partial derivative of $D_i(u_k)$ with respect to u_ℓ by

$$\frac{\partial D_i}{\partial u_\ell} = D_{i,\ell}. \tag{4.3}$$

After substitution of (4.1) and (3.28) the equation (4.2) reads as follows

$$(\sum_p A^p_{ki}\ \sigma^p_k)D_{i,\ell}\ \delta u_\ell = f_\ell \delta u_\ell . \tag{4.4}$$

The condition that (4.4) shall hold for all kinematically admissible δu_ℓ gives the equations of equilibrium for the structure

$$(\sum_p A^p_{ki}\ \sigma^p_k)D_{i,\ell} = f_\ell . \tag{4.5}$$

We have seen that in the presence of creep and plastic deformation σ^p_k can only be determined from rate equations, while further $D_{i,\ell}$ in general a complicated non-linear function is of the coordinates u_k in the deformed state.

From (4.5) we derive the following incremental equations

$$\left[D_{i,k}\ \{\sum_p A^p_{mi}(S^p_{mn}-Y^p_{mn})A^p_{nj}\}\ D_{j,\ell} + (\sum_p A_{mi}\ \sigma^p_m)\ D_{i,k\ell}\right]\ \Delta u_\ell =$$

$$= \Delta f_k + D_{i,k}\ \{\sum_p A^p_{mi}\ (S^p_{mn}-Y^p_{mn})\ c^p_n\}\ \Delta t. \tag{4.6}$$

The matrix coefficients Y^p_{mn} and the vector components c^p_n depend on the stress parameters σ^p_m as indicated in (3.15) and (3.16). Once the increments of Δu_k have been determined the increments of σ^p_m follow from

$$\Delta\sigma^p_m = (S^p_{mn}-Y^p_{mn})(A^p_{ni}\ D_{i,k}\ \Delta u_k - c^p_n\ \Delta t). \tag{4.7}$$

In shorthand notation the equation (4.6) can be written in the form

$$(K_{ij} - P_{ij} + G_{ij})\ \Delta u_j = \Delta f_j + a_j\ \Delta t. \tag{4.8}$$

Here the matrix coefficients P_{ij} and the vector elements a_j represent the physical non-linearity due to creep and plasticity. The geometrical non-linearities are contained in the quantities $D_{i,k}$ and $\sigma_i D_{i,k\ell} = G_{k\ell}$.

Numerical integration of the structural equations for prescribed increments of f_j with the aid of a computer algorithm meets with no serious difficulties as long as the expression

$$\delta u_i (K_{ij} - P_{ij} + G_{ij})\delta u_j$$

is positive definite. This implies that for all kinematically admissible δu_i with the associated δf_j must hold

$$\delta f_j \delta u_j = \delta u_i \, (K_{ij} - P_{ij} + G_{ij}) \delta u_j > 0. \tag{4.9}$$

This is a necessary and sufficient condition for stability of the structure in the given state of deformation and loading. We observe that

$$\delta u_i (K_{ij} - P_{ij}) \, \delta u_j > 0 \tag{4.10}$$

is the condition for stable material behaviour, while

$$\delta u_i G_{ij} \, \delta u_j > 0 \tag{4.11.a}$$

implies a geometric stabilization, and

$$\delta u_i G_{ij} \, \delta u_j < 0 \tag{4.11.b}$$

a geometric destabilization.

When in the course of a loading program the matrix $[K - P + G]$ becomes singular, it means that collapse of the structure will take place. Determination of the collapse load in proportional loading by numerical integration of the equations (4.6) is usually more efficient than by application of the first or second collapse theorem. Even in the case of plane frameworks in the formulation of a linear programming algebra problem the determination of the collapse load may require more computer time than a complete elastic-plastic analysis of the structure, which on the basis of geometrically linearized equations leads to the same collapse load.

5. Some remarks on the applications in engineering

If the material of a structure meets the prime requirement, that of ductility, plasticity theory enables us to calculate the strength of the structure in terms of collapse loads. This is a more realistic approach than an assessment of the strength of the structure on the basis of the elastic stress distribution, calculated for the working load. Variable factors, settlements, imperfect fabrication, stresses induced on manufacture and erection may make the elastic stress distribution a virtually meaningless estimate of the actual state of stress. Of course it can serve as a reasonable equilibrium state for the structure on which to base a safe estimate of its strength. How safe the structure is with respect to the highest loads, that it has to withstand, can only be ascertained by a plastic analysis. This analysis has therefore justifiably found its way into the building codes.

While the elastic stress distribution in itsself may be virtually meaningless, after shakedown to time independent plastic strains it determines accurately the range of stress under fluctuating loads, and is as such of significance for low and high cycle fatigue.

In the presence of creep, and in a more discontinuous fashion due to plastic deformation, a structure has the property of fading memory. The response to a sequence of loads damps out in the sense that its effect on the stress distribution is completely superseded by subsequent loads.

In the creep range, that is at temperatures close to half the melting temperature and higher, the elastic stress distribution of metal structures is again of interest only for the fluctuating part of the load. For loads of long duration the elastic effects may be neglected altogether in creep analysis. Like the limiting states of plasticity the asymptotic states of stress and deformation rate in creep are independent of the order and speed of load-application.

If the structure has to meet certain conditions of limited deformation, creep and plasticity theory can be applied to ensure that these conditions are being fulfilled. The conditions for creep and plastic deformation are reasonably well understood, but not those for fracture. We can only describe qualitatively the factors that delay fracture and promote continued deformation. The superposition of a hydrostatic pressure on a simple tensile test can counteract the tensile stresses set up in the "neck" of the specimen and promote greater plastic flow. On the other hand, the presence of flaws, cracks or notches in a member under load may lead to high triaxial tensile stresses set up and under certain conditions a catastrophic brittle failure may result.

Fracture mechanics is an active field of research nowadays. The information it requires can be classified in three categories: the material properties, the possible flaws and defects and the stress distribution in the structure. Computer methods for creep and plasticity analysis can furnish the stress distribution, provided the model of material behaviour being used gives a sufficiently accurate description of the actual behaviour, such as it can only be ascertained in careful testing.

References

1. 'Espinasse, Margaret: Robert Hooke, p. 70, London, William Heinemann Ltd., 1956.

2. Vicat, M.: Note sur l'allongement progressif du fil de fer soumis a diverses tensions, Annales de Chemie et de Physique, 54, 35, (1833); Ueber die fortschreitende Verlängerung eines Metalldrahts unter der Wirkung von Zugkräften, Poggendorf's Annalen der Physik, 1, 108, (1834).

3. Livesley, R.K.: Linear programming in structural analysis and design, International symp. on computers in optimisation of structural design, Dept. of Civil Eng., Univ. of Wales, Swansea, (1972).

4. Cohn, M.Z., Ghosh, S.K. and Parimi, S.R.: Unified approach to theory of plastic structures, J. Eng. Mech. Div., Proc. A.S.C.E., EM5, pp. 1133-1157, (1972).

5. Vajda, S.: Mathematical programming, London, Addison-Wesley publ. Co., inc., 1961.

6. Prager, W.: An introduction to Plasticity, Reading, Mass., U.S.A., Addison-Wesley publ. co. inc., 1959.

7. Wartenberg, H.V.: Ueber elastische Nachwirkung bei Metallen, Deutsche Phys. Gesellschaft Berichte, 20, 113, (1918).

8. Besseling, J.F.: A theory of elastic, plastic and creep deformations of an initially isotropic material, J. Appl. Mech., 25, 529, (1958).

CREEP IN CONTINUA AND STRUCTURES

J. HULT, Gothenburg

1. Introduction

Creep mechanics is a young branch of solid mechanics. Even though its foundations were established by experimental and theoretical studies early in this century, creep phenomena came to gain engineering importance only in the last few decades.

This development is reflected in the rate of appearance of monographs on creep and in their main contents. The works of NORTON /1/, TAPSELL /2/ and SULLY /3/ deal almost entirely with methods of creep testing and with creep properties of specific engineering materials. The concept of "limiting creep stress", being a stress below which no creep deformation occurs, is discussed at some length. In contrast to fatigue, where a well defined endurance limit often exists, no such creep limit has been found even in tests conducted over extremely long periods of time. These early monographs, all written before 1950, deal with creep as a material property, but give no advice about design of components and structures.

Such engineering aspects dominate the works of MALININ /4/, FINNIE and HELLER /5/, KACHANOV /6/, ODQVIST and HULT /7/, ODQVIST /8/, HULT /9/, RABOTNOV /10/ and PENNY and

---

Some results reported here were obtained during research supported by the Swedish Board for Technical Development.

MARRIOTT /11/. Except /4/ these were all written in the last fifteen years. Several international conferences have also been devoted to engineering creep problems, such as those in Stanford 1960 /12/, New York-London 1963 /13/, London 1964 /14/, Gothenburg 1970 /15/ and Philadelphia 1973-London 1974 /16/. The A. E. JOHNSON memorial volume /17/ is also a significant source of information on recent results in creep mechanics, as well as review articles by ODQVIST /18/, HOFF /19/, JOHNSON /20/, FINNIE /21/, ELLISON /22/ and BORESI and SIDEBOTTOM /23/.

After the basic characteristics of creep phenomena had been discovered, methods were soon developed to calculate certain quantities of importance in design work. Formulas were derived for maximum stress and for rate of deformation in certain common elements such as beams under bending, cylinders under pressure and turbine disks under rotation.

This was before the advent of the electronic computer, and consequently much effort was spent to develop simple approximate formulas to be used in design work. The complexity of creep behaviour yet limited these analyses to rather simple structures and load. Present day availability of and ready access to computers has changed all this. Calculations of stress fields and deformation histories may now be made for almost any type of structure and loading program. This new state of the art is clearly brought out in /7/.

One result of this recent development is that creep mechanics research has turned more to problems of creep rupture, where theory is not as well developed. In particular rational design rules against brittle creep rupture are almost non existent. This is a severe deficiency on account of the insidious character of a brittle rupture. In contrast to ductile creep rupture this occurs without any visible advance warning. Some recent developments of rupture analyses will be presented in the sequel. A review will also be given of various methods of analysis of stable creep deformation.

2. Basic laws of deformation and damage

Deformation and damage are two quantities describing changes occurring in a body subject to stress. For a bar under tensile load the longitudinal strain ε is a relevant measure of deformation. Damage may be visualized as an interior creation of voids and other cavities, which decrease the load carrying are from A to ψA. The quantity ψ was introduced by KACHANOV /24/ and denoted material 'continuity'. The quantity

$$\omega = 1 - \psi \tag{2.1}$$

is then a direct measure of the inner deterioration of the material. It will be denoted the damage. For a completely intact material $\omega=0$, whereas complete destruction corresponds to $\omega=1$.

If σ denotes the nominal stress in the bar, the average net stress transmitted by the load carrying area is σ/ψ. This net stress will be denoted s, and from eq. (2.1) then follows

$$s = \frac{\sigma}{1-\omega} \tag{2.2}$$

as defined by RABOTNOV /25/.

Stationary states of creep deformation and creep damage will be defined as states where the rates of increase of deformation and damage at constant net stress depend only on this stress, i.e.

$$\frac{d\varepsilon}{dt} = F(s) \tag{2.3}$$

$$\frac{d\omega}{dt} = f(s) \tag{2.4}$$

For instantaneous increases in net stress the following relations are postulated

$$\frac{d\varepsilon}{ds} = G'(s) \tag{2.5}$$

$$\frac{d\omega}{ds} = g'(s) \tag{2.6}$$

Assuming strain and damage to be zero in the unstressed state eq:s (2.5) and (2.6) integrate to

$$\varepsilon = G(s) \tag{2.7}$$

$$\omega = g(s) \tag{2.8}$$

For arbitrarily increasing stress s eq:s (2.3) through (2.6) combine to give the general laws of deformation and damage

$$\frac{d\varepsilon}{dt} = G'(s)\,\frac{ds}{dt} + F(s) \tag{2.9}$$

$$\frac{d\omega}{dt} = g'(s)\,\frac{ds}{dt} + f(s) \tag{2.10}$$

These relations were postulated by BROBERG /26/ as generalizations of earlier forms suggested by HULT /27/ and HULT and BROBERG /28/. If no damage occurs, i.e. $\omega \equiv 0$, then $s \equiv \sigma$ and eq. (2.9) takes the form

$$\frac{d\varepsilon}{dt} = G'(\sigma)\,\frac{d\sigma}{dt} + F(\sigma) \tag{2.11}$$

as previously suggested by ODQVIST /29/. Since damage creation is an irreversible process, eq. (2.10) has to be modified for cases of alternating stress, cf. HULT /30/. For quite general cases of prescribed cyclic stressing CHRZANOWSKI /31/ derived several features characteristic of creep-fatigue interaction as consequences of the damage law (2.10).

A somewhat similar damage law, relating ω to both s and $\dot{\varepsilon}$, viz.

$$\frac{d\omega}{dt} = \frac{\partial\omega}{\partial s}\cdot\frac{ds}{dt} + \frac{\partial\omega}{\partial\varepsilon}\cdot\frac{d\varepsilon}{dt} \tag{2.12}$$

was proposed by EIMER /32/. A detailed comparison between eq:s (2.10) and (2.12) remains to be done.

For multiaxial states of stress the no damage deformation law (2.11) takes the form

$$\frac{d\varepsilon_{ij}}{dt} = G'(\sigma_e)\cdot\frac{3}{2}\,\frac{ds_{ij}}{dt} + F(\sigma_e)\cdot\frac{3}{2}\,\frac{s_{ij}}{\sigma_e} \tag{2.13}$$

assuming the material to be isotropic and the deformation to be isochoric, cf. ODQVIST /29/. Here σ_e denotes the von Mises effective stress and s_{ij} the stress deviation tensor. So far no attempts have been made to generalize the deformation law (2.9) to multiaxial states of stress. A major difficulty here is to devise the multiaxial counterparts of the uniaxial stress s. Likewise the damage law (2.10) has not yet been restated in multiaxial form. Rupture analyses based on the damage law

$$\frac{d\omega}{dt} = f(s) \tag{2.14}$$

and its multiaxial counterparts have been made by MARTIN and LECKIE /33/ and HAYHURST and LECKIE /34/. A deformation law of the form

$$\varepsilon^{\alpha} \frac{d\varepsilon}{dt} = F(s) \tag{2.15}$$

reproducing also the initial decelerating rate of creep, was proposed by CHRZANOWSKI /35/ and was combined with a damage law of type (2.14) to analyze various problems of creep rupture, cf. CHRZANOWSKI /36/.

The functional expressions appearing in eq:s (2.9) and (2.10) may be determined by adaption to experimental data. For many materials power type expressions such as

$$\frac{d\varepsilon}{dt} = B_0 \; s^{n_0} \frac{ds}{dt} + B \; s^{n} \tag{2.16}$$

$$\frac{d\omega}{dt} = C_0 \; s^{\nu_0} \frac{ds}{dt} + C \; s^{\nu} \tag{2.17}$$

have been found adequate.

3. Stable creep deformation, methods of analysis

In absence of damage the deformation law, usually called the creep law, for stationary creep takes the form (2.11) or (2.13). This combines with the laws of equilibrium and compatibility to the field equations for any given element or structure. On account of the usually nonlinear forms of G' and F closed form solutions are an exception, even for cases of constant loading. Formidable mathematical difficulties may arise in cases with time variable loads. Stochastic loads have been considered by PARKUS and ZEMAN /37/ and by BJÖRKENSTAM /38/ and others.

In spite of present day computer availability simplifying methods of creep analysis are still called for. They may often cancel the need for extensive numerical calculation programs and do often offer useful insight into the mechanism of the problem at hand.

3.1 Elastic analogue

When a structure under creep conditions is subjected to instantaneously applied constant forces, a set of initial stresses will arise which then gradually change. If no damage is created, and deformation is governed by the law (2.13), the initial stresses will be determined by

$$d\varepsilon_{ij} = G'(\sigma_e) \cdot \frac{3}{2}\, ds_{ij} \qquad (3.1)$$

As creep deformation proceeds, the creep strain comes to dominate more and more over the initial strain, which may finally be neglected, so that eq. (2.13) takes the form

$$\frac{d\varepsilon_{ij}}{dt} = F(\sigma_e) \cdot \frac{3}{2}\frac{s_{ij}}{\sigma_e} \qquad (3.2)$$

With the load being constant the stress field will approach a constant state and eq. (3.2) may then be formally integrated to give

$$\varepsilon_{ij} = \varepsilon_{ij}^{0} + F(\sigma_e) \cdot \frac{3}{2}\frac{s_{ij}}{\sigma_e} \cdot t \qquad (3.3)$$

where ε_{ij}^{0} is a ficticious initial strain field. Hence after sufficient time the deformation law takes the form

$$\varepsilon_{ij} = F(\sigma_e) \cdot \frac{3}{2}\frac{s_{ij}}{\sigma_e} \cdot t \qquad (3.4)$$

For constant force loading the same state of stress would arise if the deformation law were of the form

$$\varepsilon_{ij} = F(\sigma_e) \cdot \frac{3}{2}\frac{s_{ij}}{\sigma_e} \cdot C \qquad (3.5)$$

where C is a constant. Hence the stationary stress field during creep under constant forces may be found from an analogous law of time independent elastic deformation. This analogue, strictly formulated by HOFF /39/, has found standard use in design calculations, where stationary states of stress and rate of deformation are of prime importance. Detailed studies of the approximations involved in using this elastic analogue have been made by CALLADINE /40/ and PENNY and MARRIOTT /11/ and others. Conditions for stationary states of stress to be reached even under time variable loads have been examined by HULT /41/.

3.2 Reference stress methods

Analyses of stress redistributions due to creep under constant loads have shown a certain feature, common to many structures. The stress is found to be nearly constant in one small region in the structure all during the redistribution process, cf. Fig. 1.

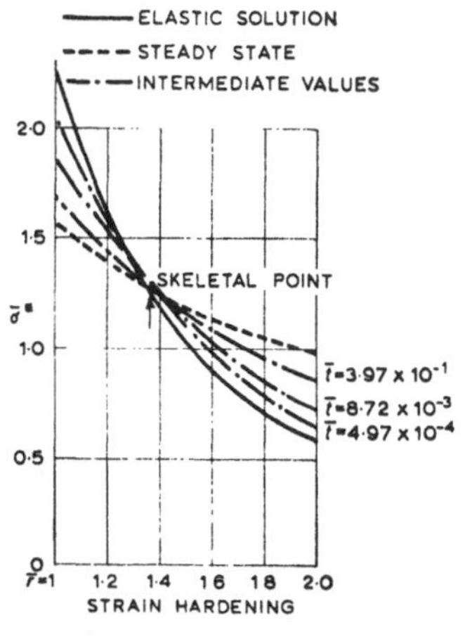

Fig. 1. Redistribution of effective stress in thick-walled cylinder under pressure, from /42/

MARRIOTT and LECKIE /42/ coined the expression 'skeletal point' for this constant stress region.

In structures with one kinematical degree of freedom the overall deformation is uniquely determined by the deformation at one point. Hence the stress at a skeletal point determines the overall deformation history during the whole creep process. If the skeletal point stress can be calculated from given loads, the deformation history may be determined from one single creep test performed at that particular stress, called the reference stress for the structure. In this way detailed knowledge of the creep law is made unnecessary.

The reference stress concept has been developed also from other aspects, cf. SODERBERG /43/, ANDERSON et al /44/, MACKENZIE /45/, SIM /46/ and JOHNSSON /47/. The reference stress technique has been applied to relaxations problems by SPENCE and HULT /48/.

3.3 Energy theorems

In solving stationary creep problems by the elastic analogue all energy theorems for nonlinear elastic deformation are available, cf. HODGE and VENKATRAMAN /49/ and several other applications.

Methods to determine upper and lower bounds on creep displacement rates have been developed by MARTIN /50/ and PALMER /51/ and by PONTER /52/ for variable loading. Such methods very often provide sufficiently accurate estimates for practical design needs.

4. Ductile instability

When a tensile bar under creep conditions is subject to constant force loading, its length will increase at an accelerating rate. This is due to the simultaneous decrease in cross sectional area. For isochoric deformation the following relation holds between load P, (real) stress σ and (natural) strain ε

$$\frac{dP}{P} = \frac{d\sigma}{\sigma} - d\varepsilon \qquad (4.1)$$

This combines with the no damage deformation law (2.11) to give

$$d\varepsilon[1-\sigma G'(\sigma)] = \sigma G'(\sigma)\,\frac{dP}{P} + F(\sigma)\,dt \qquad (4.2)$$

Hence $d\varepsilon/dP \to \infty$ during load application (t=const) or $d\varepsilon/dt \to \infty$ during the creep deformation process (P=const), when σ

reaches σ_R defined by

$$1 - \sigma_R\, G'(\sigma_R) = 0 \qquad (4.3)$$

This close connection between ductile instantaneous rupture and ductile creep rupture was first observed by CARLSON /53/. Other cases of ductile instability have been analyzed by STORÅKERS /54/.

In absence of instantaneous deformation, i.e. $G'(\sigma)\equiv 0$ in eq. (2.11), eq. (4.3) yields $\sigma_R=\infty$. This case was first analyzed by HOFF /55/, who took $F(\sigma)$ to be a power function. For an arbitrary form of $F(\sigma)$ the rupture time t_R is obtained as follows. With P=const and $G'(\sigma)\equiv 0$ eq:s (4.1) and (2.11) combine to give

$$dt = \frac{d\sigma}{\sigma F(\sigma)} \qquad (4.4)$$

and hence

$$t_R = \int_{P/A_0}^{\infty} \frac{d\sigma}{\sigma F(\sigma)} \qquad (4.5)$$

where A_0 denotes the initial cross sectional area of the bar. Eq. (4.5) gives the creep rupture time t_R corresponding to the constant tensile force P.

For variable load P(t) according to Fig. 2, a corresponding ficticious rupture time $t_R(t)$ may be calculated from eq. (4.5). Rupture occurs at time t_R^*. It is of interest to calculate the magnitude of the integral

$$J = \int_0^{t_R^*} \frac{dt}{t_R(t)} \qquad (4.6)$$

For the step loading shown in Fig. 3 the integral takes the form

$$J = \frac{T}{t_{R1}} + \frac{t_R^*-T}{t_{R2}} \qquad (4.7)$$

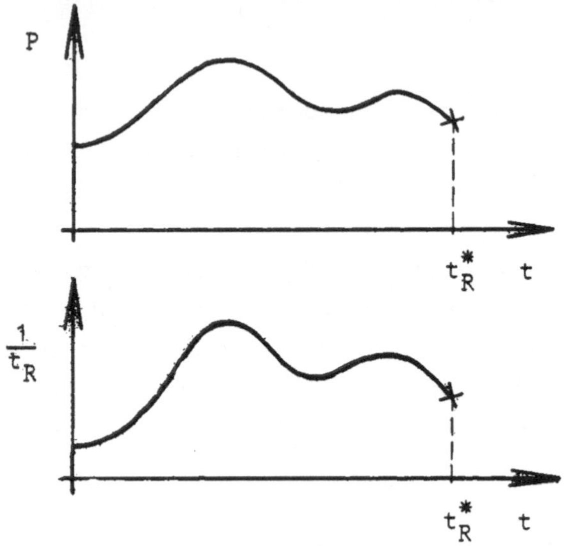

Fig. 2. Variable load P(t) and corresponding inverted rupture time $1/t_R(t)$

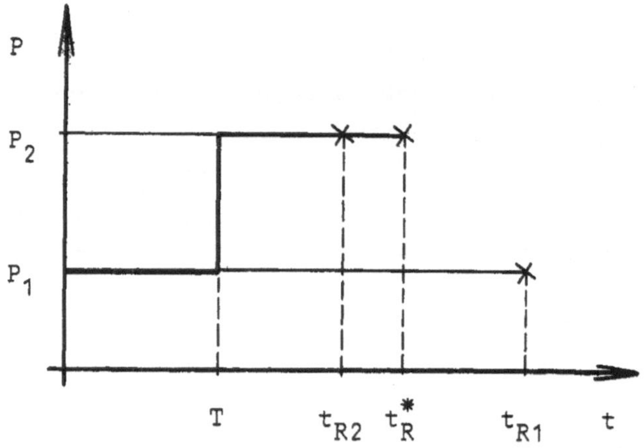

Fig. 3. Step loading leading to rupture

where

$$T = M(\sigma_1) - M(\sigma_T) \tag{4.8}$$

$$t_{R1} = M(\sigma_1) \tag{4.9}$$

$$t_{R2} = M(\sigma_2) \tag{4.10}$$

$$t_R^* - T = M(\sigma_T \cdot P_2/P_1) \tag{4.11}$$

with

$$M(x) = \int_{-x}^{\infty} \frac{d\sigma}{\sigma F(\sigma)} \tag{4.12}$$

and

$$\sigma_1 = P_1/A_0 \tag{4.13}$$

$$\sigma_2 = P_2/A_0 \tag{4.14}$$

$$\sigma_T = \sigma_1 \cdot e^{\varepsilon_T} \tag{4.15}$$

Here ε_T denotes the strain at time T. Hence eq. (4.7) takes the form

$$J = 1 - \frac{M(\sigma_1 e^{\varepsilon_T})}{M(\sigma_1)} + \frac{M(\sigma_2 e^{\varepsilon_T})}{M(\sigma_2)} \tag{4.16}$$

from which it follows that J=1, independent of σ_1, σ_2 and ε_T, if and only if M(x) is a power function. Since, from eq. (4.12)

$$F(x) = - \frac{1}{xM'(x)} \tag{4.17}$$

it then follows that the life fraction rule

$$\int_0^{t_R^*} \frac{dt}{t_R(t)} = 1 \tag{4.18}$$

is fulfilled if and only if F(σ) in eq. (2.11) is a power function. In a doubly logarithmic plot of dε/dt versus σ a power function F(σ) corresponds to a straight line, cf. Fig. 4. Such a material is usually denoted a Norton material, cf. /1/. If the curve is concave upwards, the material is said to be of hypo-Norton type, if concave downwards it is of hyper-Norton type. For step loading the following results for the integral (4.6) then follow from eq:s (4.7) through (4.15)

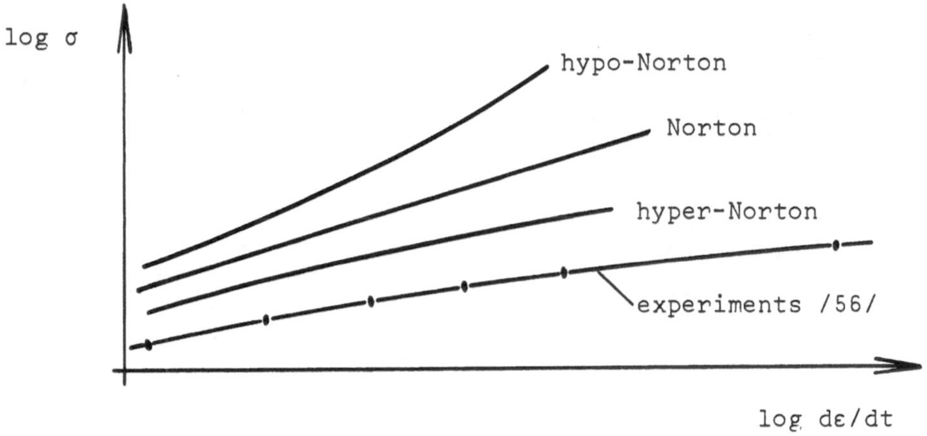

Fig. 4. Creep rate diagram

	Step up loading	Step down loading
Hypo-Norton	J>1	J<1
Hyper-Norton	J<1	J>1

These conclusions may be compared with experimental results for an aluminum alloy at 180°C, obtained by MARRIOTT and PENNY /56/. A log-log plot of $d\varepsilon/dt$ versus σ, taken at $t=100$ hours, is given in Fig. 4. It shows the material to be of hyper-Norton type. The following results were obtained in six one step load tests:

σ_1(ksi)	σ_2(ksi)	J	step type
18	20	0.8	up
20	18	1.2	down
20	14	1.2	down
14	20	0.6	up
20	14	1.3	down
24	18	1.3	down

Hence complete qualitative agreement is obtained with the theoretical predictions.

5. Brittle instability

Conditions for purely brittle rupture may be derived directly from the damage law (2.10). With s according to eq. (2.2) follows, since $\sigma=P/A$ where A now is constant

$$d\omega\left[1 - \frac{\sigma}{(1-\omega)^2}\, g'\left(\frac{\sigma}{1-\omega}\right)\right] = \frac{\sigma}{1-\omega}\, g'\left(\frac{\sigma}{1-\omega}\right)\frac{d\sigma}{\sigma} + f\left(\frac{\sigma}{1-\omega}\right) dt \tag{5.1}$$

Hence $d\omega/d\sigma\to\infty$ during load application (t=const) or $d\omega/dt\to\infty$ during the creep deterioration process (σ=const), when the following relation between σ and ω is satisfied

$$1 - \frac{\sigma}{(1-\omega)^2}\, g'\left(\frac{\sigma}{1-\omega}\right) = 0 \tag{5.2}$$

Such brittle instability may arise during load application, in which case, from eq. (2.8)

$$\omega = g\left(\frac{\sigma}{1-\omega}\right) \tag{5.3}$$

or during creep at constant stress

$$\sigma = \sigma_0 \tag{5.4}$$

From eq:s (5.2) and (5.3) follow σ_R and ω_R as shown in Fig. 5; from eq:s (5.2) and (5.4) follows ω_*, where

$$\omega_R < \omega_* < 1 \tag{5.5}$$

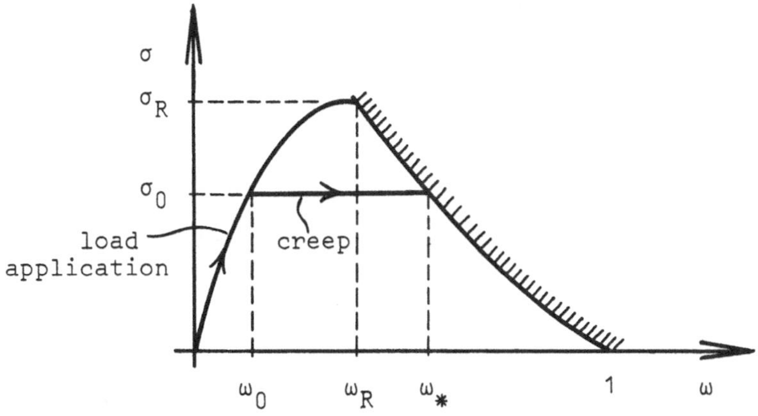

Fig. 5. Brittle instability locus

Previous analyses of brittle creep rupture have been based on the damage law (2.4) rather than (2.10). Instability in the sense $d\omega \to \infty$ is then never attained, and the rupture condition is therefore taken as $\omega=1$. The present analysis, due to BROBERG /26/, shows that inclusion of the time independent damage term g(s) implies that an unstable state may be reached already during load application (at ω_R), and that it will always be reached (at ω_*) before complete destruction has taken place.

In absence of g(s) the life fraction rule is found to hold for purely brittle rupture if and only if f(s) in eq. (2.10) is a power function, as was originally assumed by KACHANOV /24/, cf. /57/.

6. Mixed type instability

Assuming both ductile isochoric deformation and brittle deterioration to take place eq:s (2.9), (2.10) and (4.1) combine to show the simultaneous occurrence of ductile instability ($d\varepsilon \to \infty$) and brittle instability ($d\omega \to \infty$) when

$$1 - \frac{s}{1-\omega} g'(s) - sG'(s) = 0 \qquad (6.1)$$

Comparison with the instability criteria (4.3) and (5.2) shows that mixed type instability will occur at a lower stress and a smaller damage than in the two pure cases, cf. /26/ and /58/. Further studies of conditions for instability, in particular under cyclic and other time variable loads are under way.

References

1. NORTON, F.H.: Creep of Steel at High Temperatures. New York: McGraw-Hill. 1929.

2. TAPSELL, H.J.: Creep of Metals. Oxford University Press. 1931.

3. SULLY, A.H.: Metallic Creep and Creep Resistant Alloys. London: Butterworths. 1949.

4. MALININ, N.N.: Fundamentals of Creep Calculations (in Russian). Moscow: Mazhgiz. 1948.

5. FINNIE, I. and W.R. HELLER: Creep of Engineering Materials. New York: McGraw-Hill. 1959.

6. KACHANOV, L.M.: Theory of Creep. Moscow: Gos.Izdat.Fis.-Mat.Lit. 1960.

7. ODQVIST, F.K.G. and J. HULT: Kriechfestigkeit metallischer Werkstoffe. Berlin: Springer. 1962.

8. ODQVIST, F.K.G.: Mathematical Theory of Creep and Creep Rupture. Oxford University Press. 1966.

9. HULT, J.: Creep in Engineering Structures. Waltham, Mass: Blaisdell. 1966.

10. RABOTNOV, Yu.N.: Creep Problems in Structural Members (in Russian 1966). Amsterdam: North-Holland. 1969.

11. PENNY, R.K. and D.L. MARRIOTT: Design for Creep. London: McGraw-Hill. 1971.

12. IUTAM Colloquium, Creep in Structures, Stanford 1960. Proc. (ed. N.J.HOFF). Berlin: Springer. 1962.

13. ASME-ASTM-IME Joint International Conference on Creep, New York-London 1963. Proc. (ed. IME). London: Institution of Mechanical Engineers. 1964.

14. IME Conference, Thermal Loading and Creep in Structures and Components, London 1964. Proc. (ed. IME). London: Institution of Mechanical Engineers. 1964.

15. IUTAM Symposium, Creep in Structures, Gothenburg 1970. Proc. (ed. J.HULT). Berlin: Springer. 1972.

16. IME Conference, Creep and Fatigue in Elevated Temperature Applications, Sheffield 1974. Proc. to be published by IME. London: Institution of Mechanical Engineers.

17. Advances in Creep Design - the A.E. JOHNSON Memorial Volume (ed. A.I.SMITH and A.M.NICHOLSON). London: Applied Science Publishers. 1971.

18. ODQVIST, F.K.G.: Advances in Theories of Creep of Engineering Materials. Appl. Mech. Rev. 7, 517-519 (1954).

19. HOFF, N.J.: A Survey of the Theories of Creep Buckling. Proc. 3 US Nat. Congr. Appl. Mech. 1958, 22-49. New York: ASME. 1958.

20. JOHNSON, A.E.: Complex-Stress Creep of Metals. Mat. Rev. 5, 447-506 (1960).

21. FINNIE, I.: Stress Analysis for Creep and Creep Rupture. Appl. Mech. Rev. 13, 373-387 (1960).

22. ELLISON, E.G.: A Review of the Interaction of Creep and Fatigue. J.Mech.Eng.Sci. 11, 318-339 (1969).

23. BORESI, A.P. and O.M. SIDEBOTTOM: Creep of Metals under Nultiaxial States of Stress. Nuclear Engng and Design 18, 415-456 (1972).

24. KACHANOV, L.M.: Time of the Rupture Process under Creep Conditions. Izv. Akad. Nauk SSSR, Otd. Tekh. Nauk. Nr. 8, 26-31 (1958).

25. RABOTNOV, Yu.N.: Creep Rupture. Proc. XII Int. Congr. Appl. Mech., Stanford 1968, 342-349 (Ed. M.HETÉNYI and W.G.VINCENTI). Berlin: Springer. 1969.

26. BROBERG, H.: A new Criterion for Brittle Creep Rupture. To appear in J.Appl.Mech.

27. HULT, J.: Creep Strength of Structures. Udine: CISM. 1973.

28. HULT, J. and H. BROBERG: Creep Rupture under Cyclic Loading. Bulgarian 2 Nat. Congr. Theor. Appl. Mech., Varna 1973.

29. ODQVIST, F.K.G.: Influence of Primary Creep on Stresses in Structural Parts. VIII Int. Congr. Appl. Mech., Istanbul 1952. Trans. Roy. Inst. Tech. Stockholm, Nr. 66 (1953).

30. HULT, J.: Structural Creep Behaviour under Alternating Load. Paper to be presented at /16/.

31. CHRZANOWSKI, M.: Use of the Damage Concept in Describing Creep-Fatigue Interaction under Prescribed Stress. Chalmers University, Gothenburg (internal report), 1974.

32. EIMER, C.: Rheological Approach to Problems of Cumulative Damage and Strength. Bull. Acad. Pol. Sci., Ser. Sci. Tech. 20, 255-261 (1972).

33. MARTIN, J.B. and F.A. LECKIE: On the Creep Rupture of Structures. J. Mech. Phys. Solids 20, 223-238 (1972).

34. HAYHURST, D.R. and F.A. LECKIE: The Effect of Creep Constitutive and Damage Relationships upon the Rupture Time of a Solid Circular Torsion Bar. J. Mech. Phys. Solids 21, 431-446 (1973).

35. CHRZANOWSKI, M.: On the Possibility of Describing the Complete Process of Metallic Creep. Bull. Acad. Pol. Sci., Ser. Sci. Tech. 20, 75-81 (1972).

36. CHRZANOWSKI, M.: The Description of Metallic Creep in the Light of Damage Hypothesis and Strain Hardening. Diss. hab., Politechnika Krakowska, Kraków (1973).

37. PARKUS, H. and J.L. ZEMAN: Some Stochastic Problems of Thermoviscoelasticity. Proc. IUTAM Symposium, Thermoinelasticity, East Kilbride 1968 (ed. B.A.BOLEY), 226-240. Wien: Springer. 1970.

38. BJÖRKENSTAM, U.: Random Loading on Structures under Creep. Diss. Chalmers University, Gothenburg 1973.

39. HOFF, N.J.: Approximate Analysis of Structures in the Presence of Moderately Large Creep Deformations. Quart. Appl. Math. 12, 49-55 (1954).

40. CALLADINE, C.R.: Time-Scales for Redistribution of Stress in creep of Structures. Proc. Roy. Soc. A. 309, 363-375 (1969).

41. HULT, J.: On the Stationarity of Stress and Strain Distributions in Creep. Proc. IUTAM Symposium, Second-Order Effects in Elasticity, Plasticity and Fluid Dynamics (ed. M.REINER and D.ABIR), 352-361. New York: Macmillan. 1964.

42. MARRIOTT, D.L. and F.A. LECKIE: Some Observations on the Deflections of Structures during Creep. Ref. /14/, 115-125.

43. SODERBERG, C.R.: Interpretation of Creep Tests on Tubes. Trans. ASME 63, 737-740 (1941).

44. ANDERSON, R.G., L.R.T. GARDNER and W.R. HODGKINS: Deformation of Uniformly Loaded Beams obeying Complex Creep Laws. J. Mech. Eng. Sci. 5, 238-244 (1963).

45. MACKENZIE, A.C.: On the Use of a Single Uniaxial Test to Estimate Deformation Rates in some Structures Undergoing Creep. Int. J. Mech. Sci. 10, 441-453 (1968).

46. SIM, R.G.: Reference Stress Concepts in the Analysis of Structures during Creep. Int. J. Mech. Sci. 12, 561-573 (1970).

47. JOHNSSON, A.: The Reference Stress Method in Creep Design. Diss. Chalmers University, Gothenburg 1973.

48. SPENCE, J. and J. HULT: Simple Approximations for Creep Relaxation. Int. J. Mech. Sci. 15, 741-755 (1973).

49. HODGE, P.G. Jr. and B. VENKATRAMAN: Approximate Solutions on Some Problems in Steady Creep. Proc. Symposium su la Plasticita nella Scienza delle Costruzioni, Varenna 1956.

50. MARTIN, J.B.: A Note on the Determination of an Upper Bound on Displacement Rates for Steady Creep Problems. J. Appl. Mech. 33, 216-217 (1966).

51. PALMER, A.C.: A Lower Bound on Displacement Rates in Steady Creep. J. Appl. Mech. 34, 216-217 (1967).

52. PONTER, A.R.S.: On the Stress Analysis of Creeping Structures Subject to Variable Loading. To be published in J. Appl. Mech.

53. CARLSON, R.L.: Creep-Induced Tensile Instability. J.Mech. Eng. Sci. 7, 228-229 (1965).

54. STORÅKERS, B.: Bifurcation and Instability Modes in Thick-Walled Viscoplastic Pressure Vessels. Proc. IUTAM Symposium, Creep in Structures, Gothenburg 1970 (ed. J.HULT), 333-344. Berlin: Springer. 1972.

55. HOFF, N.J.: Necking and Rupture of Rods under Tensile Loads. J. Appl. Mech. 20, 105-108 (1953).

56. MARRIOTT, D.L. and R.K. PENNY: Strain Accumulation and Rupture during Creep under Variable Uniaxial Tensile Loading. J. Strain Analysis 8, 151-159 (1973).

57. ODQVIST, F.K.G. and J. HULT: Some Aspects of Creep Rupture. Arkiv för Fysik 19:26, 379-382 (1961).

58. BOSTRÖM, P.O., H. BROBERG, L. BRÅTHE and M. CHRZANOWSKI: On Failure Conditions in Visco-Elastic Media and Structures. To be presented at IUTAM Symposium, Viscoelastic Media and Bodies, Gothenburg 1974.

WEAKENING OF ELASTIC SOLIDS BY DOUBLY-PERIODIC ARRAYS OF CRACKS

W. R. DELAMETER, AND G. HERRMANN
STANFORD, CALIFORNIA

1. Introduction

Problems concerning elastic bodies which contain multiple cracks are of obvious importance because the structural behavior of solids is strongly influenced by the presence of cracks. Solids containing cracks are weaker mechanically, not merely because at certain levels of applied loads the concentrated stresses at the crack tips may cause the crack to propagate and lead to fracture, but also because at lower stress levels the magnitudes of the elastic constants of the material can be significantly reduced by the presence of the cracks. Further, if the cracks are arranged in a preferred orientation, the elastic response of the body can be highly anisotropic.

To study the interaction of a large number of closely spaced cracks, the realistic nature of arbitrarily distributed cracks of various scales must be sacrificed to render the problem manageable from a mathematical standpoint. Therefore, problems where cracks are of the same length and arranged in periodic arrays are investigated.

Several such arrays of cracks have been previously considered. The problem of a periodic array of collinear cracks (which we shall refer to as a "row" of cracks) was solved in closed form by Koiter [1] and was later discussed by Paris and Sih [2]. What we shall call a "stack" of cracks, an array of parallel cracks which are perpendicular to the line joining their midpoints, has also been studied, and the problem is solved in closed form for the case of anti-plane strain by Louat [3]. Approximate solutions for a stack of cracks have been presented for the case of in-plane shear by Koiter [4] and for the case of tension by Ichikawa, Ohashi, and Yokobori [5], and also later by Benthem and Koiter [6]. However, doubly-periodic arrays of cracks, an obvious extension to this class of problems, have to our knowledge not been treated. Therefore, we shall investigate the weakening of elastic solids due to two doubly-periodic arrays: (1) a rectangular array, and (2) a diamond-shaped array, as shown in Fig. 1.

2. Formulation of the Problems and the Method of Solution

The technique of representing cracks by suitable continuous distributions of infinitesimal dislocations in well established, and pertinent references are cited by Bilby and Eshelby [7].

Three types of dislocations are defined for the purposes of this analysis, where in each case the line of the dislocation is parallel to the z-axis of a cartesian system of coordinates Oxyz . Type I and Type II dislocations are edge dislocations whose Burgers vectors are of constant magnitude and parallel to the y-axis and the x-axis, respectively, while a Type III dislocation is a screw dislocation with a constant Burgers vector parallel to the z-axis. The edge dislocations induce a state of plane strain, while the screw dislocation has associated with it a state of anti-plane strain.

Each crack is represented by simultaneous distributions of Type I, Type II, and Type III dislocations, where, for cracks parallel to the x-axis, the strength (i.e., magnitude of the Burgers vector) of the Type j dislocation at x is $D_j(x)$, $j = I, II, III$. These dislocation distribution functions are related to the displacement discontinuities Δu_x , Δu_y , and Δu_z across each crack by the equations

$$D_I(x) = -\frac{d}{dx}\Delta u_y(x) \tag{1a}$$

$$D_{II}(x) = -\frac{d}{dx}\Delta u_x(x) \tag{1b}$$

$$D_{III}(x) = -\frac{d}{dx}\Delta u_z(x) \tag{1c}$$

The dislocation distribution functions are determined by integral equations which result from the satisfaction of the boundary conditions at the crack faces. For freely-slipping cracks, we require that the crack faces be traction-free.

Consider an elastic body containing a doubly-periodic rectangular array of cracks, as shown in Fig. 1a. If the solid is subjected to a uniform applied stress state whose components are σ_{ij}^A , and each crack is required to be traction-free, then the following integral equations result:

$$\int_{-a}^{a} D_I(x') \; [(x'-x)^{-1} + k_I(x,x')] \; dx' = \sigma_{yy}^A/A \tag{2a}$$

$$\int_{-a}^{a} D_{II}(x') \; [(x'-x)^{-1} + k_{II}(x,x')] \; dx' = \sigma_{xy}^A/A \tag{2b}$$

$$\int_{-a}^{a} D_{III}(x') \; [(x'-x)^{-1} + k_{III}(x,x')] \; dx' = \sigma_{yz}^A/2\pi\mu \tag{2c}$$

where $A = E/4\,\pi(1-\nu^2)$, E is Young's modulus, μ is the shear modulus, and ν is Poisson's ratio. $k_I(x,x')$, $k_{II}(x,x')$, and $k_{III}(x,x')$ are the non-singular parts of the kernels of the respective integral equations. For the rectangular array of cracks, they are given by

$$k_I(x,x') = -(x'-x)^{-1} + \sum_{n=-\infty}^{\infty}\Big\{2\frac{\pi}{b}\coth\Big[\frac{\pi}{b}(x'-x+nd)\Big] - \Big(\frac{\pi}{b}\Big)^2 (x'-x+nd)\operatorname{csch}^2\Big[\frac{\pi}{b}(x'-x+nd)\Big]\Big\} \tag{3a}$$

$$k_{II}(x,x') = -(x'-x)^{-1} + \sum_{n=-\infty}^{\infty}\Big(\frac{\pi}{b}\Big)^2 (x'-x+nd)\operatorname{csch}^2\Big[\frac{\pi}{b}(x'-x+nd)\Big] \tag{3b}$$

$$k_{III}(x,x') = -(x'-x)^{-1} + \sum_{n=-\infty}^{\infty}\frac{\pi}{b}\coth\Big[\frac{\pi}{b}(x'-x+nd)\Big] \tag{3c}$$

Notice that each of these expressions is non-singular, in spite of the appearance of individually singular terms.

Similarly, for a diamond-shaped array of cracks, integral equations for $D_I(x)$, $D_{II}(x)$, and $D_{III}(x)$ which are of exactly the form of (2a,b,c) are obtained, but where the kernels are different, and for the diamond-shaped array are given by

$$k_I(x,x') = -(x'-x)^{-1} + \sum_{n=-\infty}^{\infty}\Big\{2\frac{\pi}{b}\coth\Big[\frac{\pi}{b}(x'-x+nd)\Big] - \frac{\pi^2}{b^2}(x'-x+nd)\operatorname{csch}^2\Big[\frac{\pi}{b}(x'-x+nd)\Big] + 2\frac{\pi}{b}\tanh\Big[\frac{\pi}{b}\Big(x'-x+nd+\frac{d}{2}\Big)\Big] + \frac{\pi^2}{b^2}\Big(x'-x+nd+\frac{d}{2}\Big)\operatorname{sech}^2\Big[\frac{\pi}{b}\Big(x'-x+nd+\frac{d}{2}\Big)\Big]\Big\} \tag{4a}$$

$$k_{II}(x,x') = -(x'-x)^{-1} + \sum_{n=-\infty}^{\infty} \left\{ \frac{\pi^2}{b^2} (x'-x+nd) \operatorname{csch}^2\left[\frac{\pi}{b}(x'-x+nd)\right] - \frac{\pi^2}{b^2}\left(x'-x+nd+\frac{d}{2}\right) \operatorname{sech}^2\left[\frac{\pi}{b}\left(x'-x+nd+\frac{d}{2}\right)\right] \right\} \tag{4b}$$

$$k_{III}(x,x') = -(x'-x)^{-1} + \sum_{n=-\infty}^{\infty} \left\{ \frac{\pi}{b} \coth\left[\frac{\pi}{b}(x'-x+nd)\right] + \frac{\pi}{b} \tanh\left[\frac{\pi}{b}\left(x'-x+nd+\frac{d}{2}\right)\right] \right\} \tag{4c}$$

It is apparent that $D_I(x)$, $D_{II}(x)$, and $D_{III}(x)$ for both arrays will depend on the crack spacings b and d, and on the crack half-length a, as these parameters are contained in the kernels of the integral equations that determine these functions. It should also be noted that the dislocation distributions which represent a crack in the rectangular array will be different from those which represent a crack in the diamond-shaped array, as the kernels of the integral equations which determine $D_j(x)$ are different.

Due to the complexity of the kernels, closed-form solutions to (2a,b,c) cannot be determined. However, a rapidly converging series solution can be established for singular integral equations of this form by a technique which begins with the expansion of the non-singular part of the kernel in a series of Chebychev polynomials, as outlined by the authors in a previous paper [8].

The series solutions for the dislocation distribution functions are of the form

$$D_I(x) = (\sigma_{yy}^A a/\pi A)\left(a^2-x^2\right)^{-1/2} \sum_{i=1}^{\infty} B_i^I T_{2i-1}\left(\frac{x}{a}\right) \tag{5a}$$

$$D_{II}(x) = (\sigma_{xy}^A a/\pi A)\left(a^2-x^2\right)^{-1/2} \sum_{i=1}^{\infty} B_i^{II} T_{2i-1}\left(\frac{x}{a}\right) \tag{5b}$$

$$D_{III}(x) = (2\sigma_{yz}^A a/\mu)\left(a^2-x^2\right)^{-1/2} \sum_{i=1}^{\infty} B_i^{III} T_{2i-1}\left(\frac{x}{a}\right) \tag{5c}$$

where $T_j(x)$ is the jth Chebychev polynomial of the first kind. The coefficients of the series (B_i, $i = 1,2,\ldots$) are functions of the geometry (i.e., the crack pattern and spacing).

3. Stress Intensity Factors

When a crack is represented by continuous distributions of dislocations, the stress intensity factors are determined by the dislocation density near the crack tips. Bilby and Eshelby show that the stress intensity factors K_1, K_2, and K_3, corresponding to Mode 1 (opening mode), Mode 2 (sliding mode), and Mode 3 (tearing mode) crack deformation, respectively, are given for a crack of length 2a by (cf. Eq. (142) in [7])

$$K^2 = 2\pi^3 A^2 \lim_{x \to a} (a-x)\, D^2(x) \tag{6}$$

where K_1 is determined by $D_I(x)$, K_2 by $D_{II}(x)$, and K_3 by $D_{III}(x)$, and where A is replaced by $A(1-\nu)$ for the Mode 3 case.

Substituting (5a,b,c) into (6) and noting that $T_n(1) = 1$, $n = 1,2,\ldots,$ we can show that

$$K_1/K_1^{sc} = \sum_{i=1}^{\infty} B_i^{I} \tag{7a}$$

$$K_2/K_2^{sc} = \sum_{i=1}^{\infty} B_i^{II} \tag{7b}$$

$$K_3/K_3^{sc} = \sum_{i=1}^{\infty} B_i^{III} \tag{7c}$$

where the superscript "sc" refers to the value of the stress intensity factor for a single (i.e., isolated) crack in an infinite body subjected to uniform stresses.

These ratios have been evaluated for various crack spacings of both the rectangular and diamond-shaped arrays, and values are presented as functions of b/2a for several values of d/2a in Figs. 2 - 7.

If the crack spacing in the y-direction is very much larger than the crack length, the value of each stress intensity factor approaches that associated with a single row of collinear cracks. That is, when $b/2a = \infty$,

$$K_1/K_1^{sc} = K_2/K_2^{sc} = K_3/K_3^{sc} = \left(\frac{d}{\pi a}\tan\frac{\pi a}{d}\right)^{1/2} \tag{8}$$

as discussed by Paris and Sih [2].

When b/2a is small, it becomes necessary to discuss each case separately. Let us first consider K_1/K_1^{sc} for a rectangular array of cracks, as shown in Fig. 2. It can be seen that the curves for the various values of d/2a tend to merge into the same curve as b/2a gets small. In other words, K_1/K_1^{sc} is independent of d/2a when b/2a is small, and depends only on the crack spacing in the y-direction. It has been previously shown that, for a single stack of cracks (cf. Eq. (3.67) in [6]),

$$K_1/K_1^{sc} = (b/2\pi a)^{1/2} \quad \text{as} \quad b/a \to 0 \tag{9}$$

and therefore we conclude that (9) holds for all values of $d/2a$ as $b/2a \to 0$.

Since the interaction between cracks in a single stack is such that $K_1/K_1^{sc} < 1.0$ (see [5]), but that the cracks in a single row interact so that $K_1/K_1^{sc} > 1.0$, as can be seen from Eq. (8), it seems reasonable that there exists for each value of $d/2a$ a corresponding value of $b/2a$ where the two interactive effects are cancelled by each other and K_1/K_1^{sc} is equal to unity. This is true, and the relationships between $d/2a$ and $b/2a$ for several constant values of K_1/K_1^{sc} are shown in Fig. 8.

It can be seen from Fig. 8 that the distance $d-2a$ separating adjacent crack tips in the same row must be much smaller than the distance b separating adjacent crack tips in the same stack for K_1/K_1^{sc} to equal unity. We conclude that if $b = d-2a$, the influence of the stack tending to diminish K_1/K_1^{sc} dominates the influence of the row tending to magnify K_1/K_1^{sc} .

If it is assumed that fracture occurs if the stress intensity factor is greater than or equal to an experimentally determined critical value, then a body containing a rectangular array of cracks for which the point ($d/2a$, $b/2a$) lies on or below the curve $K_1/K_1^{sc} = 1.0$ in Fig. 8 will be of the same strength or stronger (i.e., will fracture at a higher level of applied stress) than a similar body containing only a single crack of the same length. This is assuming, of course, that both bodies are loaded in tension in the direction normal to the crack faces.

Now, consider K_1/K_1^{sc} for a diamond-shaped array of cracks, as shown in Fig. 3. When investigating the value of K_1/K_1^{sc} as $b/a \to 0$, it becomes necessary to distinguish between cases where "overlap" occurs between adjacent cracks, and when there is no overlap. (Crack overlap occurs when $d < 4a$, as shown in Fig. 9). If there is no overlap, the curves representing K_1/K_1^{sc} for fixed values of $d/2a$ merge with the curve for a single stack (as given by Eq. (9)) as $b/2a \to 0$, as was true for the rectangular array. However, if there is crack overlap, separate stacks of closely spaced cracks are not formed when $b/2a \to 0$, and the values of K_1/K_1^{sc} are quite different for small $b/2a$, as can be seen from Fig. 3. Crack overlap tends to greatly decrease the value of K_1/K_1^{sc} , and hence strengthen the body in Mode 1 deformation.

The effects of the geometry on the Mode 3 stress intensity factor are similar to those seen for the Mode 1 case for both the rectangular array and the diamond-shaped array, as shown in Figs. 4 and 5. For the rectangular array and the diamond-shaped array with no overlap, K_3/K_3^{sc} tends to the value associated with a single stack of cracks as $b/2a$ gets small. This value may be determined from the dislocation distribution function given by Louat [3] to be

$$K_3/K_3^{sc} = \left(\frac{b}{\pi a} \tanh \frac{\pi a}{b}\right)^{1/2} \tag{10}$$

For the case of a diamond-shaped array where there is crack overlap, K_3/K_3^{sc} is significantly less for a given value of $b/2a$ when this ratio is on the order of unity or less, as was true for the Mode 1 case.

Consider, now, the Mode 2 stress intensity factor, as shown for the rectangular and diamond-shaped arrays in Figs. 6 and 7, respectively. For both a single row and a single stack, the cracks interact such that K_2/K_2^{sc} increases with decreasing crack spacing, and therefore it would be expected that K_2/K_2^{sc} increase monotonically with increasing crack density for each of the two arrays investigated. From Fig. 6, it may be seen that this is generally true for the rectangular array; K_2/K_2^{sc} tends to increase as both $b/2a$ and $d/2a$ decrease, and as $b/2a$ gets small, the curves merge and tend to infinity along the same curve as for a single stack. It may be concluded from Koiter's discussion of a stack of cracks [4] that

$$K_2/K_2^{sc} = (6a/\pi b)^{1/2} \quad \text{as} \quad b/a \to 0 \tag{11}$$

K_2/K_2^{sc} for a diamond-shaped array also tends to the value for a single stack of cracks as $b/2a \to 0$, and tends to the value for a single row of cracks when $b/2a$ is very large. However, while the general trend is for K_2/K_2^{sc} to increase with crack density, there are regions where the value of K_2/K_2^{sc} actually diminishes with decreased crack spacing, as can be seen from Fig. 7. For example, K_2/K_2^{sc} when $b/2a = 2.0$ and $d/2a = 2.5$ is less than unity, indicating that for this array, the body is less susceptible to fracture than a similar body containing a single crack of the same length, for the Mode 2 case. This does not occur for any combination of crack spacing for the rectangular array.

4. Change in Strain Energy and Effective Elastic Constants

It is of interest to determine not only the weakening of a body due to cracks as related to fracture, but also to study how the gross elastic response of the body is altered by the presence of cracks. When a solid containing a doubly-periodic array of parallel cracks is subjected to applied stresses, the elastic response (on a scale much larger than the crack spacing) will appear uniform but anisotropic. For example, if a tensile stress is applied in the direction parallel to the crack faces, the apparent elastic constants will be the same as if the cracks were not present. If, however, a tensile stress is applied in the direction normal to the crack faces, the values of Young's modulus and Poisson's ratio will appear to be less than the actual values of those constants for the material. Since additional strain energy due to the cracks is present, more work is done on the body by the applied loads, and hence greater surface displacements occur than would if the cracks were not present.

Consider an orthotropic solid with the following stress-strain relations:

$$\begin{aligned}
\epsilon_{xx} &= \frac{1}{E_1}\sigma_{xx} - \frac{\nu_{21}}{E_2}\sigma_{yy} - \frac{\nu_{31}}{E_3}\sigma_{zz} \\
e_{yy} &= -\frac{\nu_{12}}{E_1}\sigma_{xx} + \frac{1}{E_2}\sigma_{yy} - \frac{\nu_{32}}{E_3}\sigma_{zz} \\
\epsilon_{zz} &= -\frac{\nu_{13}}{E_1}\sigma_{xx} - \frac{\nu_{23}}{E_2}\sigma_{yy} + \frac{1}{E_3}\sigma_{zz} \\
\epsilon_{xy} &= \frac{1}{2\mu_{12}}\sigma_{xy}\ ; \quad \epsilon_{yz} = \frac{1}{2\mu_{23}}\sigma_{yz}\ ; \quad \epsilon_{xz} = \frac{1}{2\mu_{13}}\sigma_{xz}
\end{aligned} \tag{12}$$

where

$$\nu_{12}/E_1 = \nu_{21}/E_2\ , \quad \nu_{13}/E_1 = \nu_{31}/E_3\ , \quad \text{and} \quad \nu_{23}/E_2 = \nu_{32}/E_3$$

If this orthotropic solid is subjected to the uniform applied components of stress σ^A_{xz} and σ^A_{yz}, then the resulting deformation is a state of anti-plane strain, and the strain energy density W is

$$W = \frac{1}{2\mu_{13}}\left(\sigma^A_{xz}\right)^2 + \frac{1}{2\mu_{23}}\left(\sigma^A_{yz}\right)^2 \tag{13}$$

Now, consider a body of isotropic material (elastic constants E and ν) containing either a rectangular array or a diamond-shaped array of cracks and subjected to the same applied stresses. The resulting average strain energy density W^* is

$$W^* = \frac{1}{2\mu}\left(\sigma^A_{xz}\right)^2 + \frac{1}{2\mu}\left(\sigma^A_{yz}\right)^2 + \Delta E_{III}/\text{Area} \tag{14}$$

where ΔE_{III} is the change in strain energy (per unit thickness of material in the z-direction) associated with each crack, and "Area" is the area of the cell occupied by each crack, where Area = bd for the rectangular array, and Area = bd/2 for the diamond-shaped array. This last term represents the change in the average strain energy density due to the cracks.

The change in strain energy per crack for the Mode 3 case is given by the expression

$$\Delta E_{III} = \pi a^2 \left(\sigma^A_{yz}\right)^2 B_1^{III}/2\mu \tag{15}$$

where B_1^{III}, the first coefficient in the series expression for $D_{III}(x)$ in Eq. (5c), is a function of the geometry and is equal to the ratio of the change in strain energy per crack to the change in strain energy due to a single crack in an infinite body for the Mode 3 case [8]. Notice that the change in strain energy due to the cracks is independent of σ^A_{xz} since this component of stress is unaltered by cracks parallel to the x-axis.

Therefore, substituting (15) into (14), we see that

$$W^* = \frac{1}{2\mu}\left(\sigma^A_{xz}\right)^2 + \frac{1}{2\mu}\left(\sigma^A_{yz}\right)^2 \left(1 + \pi a^2 B_1^{III}/\text{Area}\right) \tag{16}$$

If the expressions for W (for a continuous orthotropic material) and W^* (for an isotropic material with cracks) are compared, we see that the strain energy density (and hence the elastic response) of each will be the same if

$$\mu_{13} = \mu \tag{17a}$$

$$\mu_{23} = \mu / \left[1 + \pi a^2 B_1^{III}/\text{Area}\right] \tag{17b}$$

In other words, (17a,b) represent the effective elastic constants of the material with cracks in a state of anti-plane strain.

By a similar process of comparing the strain energy density of an orthotropic material and the average strain energy density of an isotropic material containing cracks where both bodies are subjected to uniform stresses resulting in a state of plane strain, it may be shown that the effective elastic constants of the material with cracks are

$$\nu_{12} = \nu_{13} = \nu_{31} = \nu_{32} = \nu \tag{18a}$$

$$E_1 = E_3 = E \tag{18b}$$

$$E_2/E = \nu_{23}/\nu = \nu_{21}/\nu = 1/\ [1 + 2\pi a^2(1-\nu^2)\ B_1^I/\text{Area}] \tag{18c}$$

$$\mu_{12}/\mu = 1/\ [1 + \pi a^2(1-\nu)\ B_1^{II}/\text{Area}] \tag{18d}$$

where B_1^I and B_1^{II} are the first coefficients in the series expressions for $D_I(x)$ and $D_{II}(x)$ in (5a,b). B_1^I and B_1^{II} have been shown to be equal to the ratio of the change in strain energy per crack to the change in energy due to an isolated crack for the Mode 1 and the Mode 2 cases, respectively [8].

μ_{23}/μ is plotted as a function of $2a/b$ for several values of $2a/d$ for a rectangular array of cracks in Fig. 10 and for a diamond-shaped array in Fig. 11.

For the sake of possible comparison with experimental data, E_2/E and μ_{12}/μ shall be determined for a state of plane stress, since any experiment would be more easily performed using a thin sheet of material with slits in it. The effective elastic constants for a thin sheet containing a doubly-periodic array of cracks are

$$E_2/E = \nu_{21}/\nu = 1/\ [1 + 2\pi a^2\ B_1^I/\text{Area}] \tag{19a}$$

$$\mu_{12}/\mu = 1/\ [1 + \pi a^2\ B_1^{II}/(1+\nu)\text{Area}] \tag{19b}$$

E_2/E $(= \nu_{21}/\nu)$ for the rectangular and diamond-shaped arrays has been plotted as a function of $2a/b$ for several values of $2a/d$ in Figs. 12 and 13, respectively. Similarly, μ_{12}/μ is plotted in Figs. 14 and 15, where ν has been assigned a value of 0.3 .

From Figs. 12 and 13, it is observed that E_2/E tends to certain non-zero values as $2a/b \to \infty$. To determine what these limits are, let us examine B_1^I as $b/a \to 0$. As was the case for K_1/K_1^{sc} , B_1^I $(= \Delta E_I/\Delta E_I^{sc})$ for a rectangular array and for a diamond-shaped array with no crack overlap is independent of d/a and tends to the value of B_1^I associated with a single stack of cracks when $b/a \to 0$. It has been shown for a stack of cracks (and hence for a rectangular array and a diamond-shaped array with no overlap) that [8]

$$\lim_{b/a\to 0} \frac{\pi B_1^I}{b/a} = 1 \tag{20}$$

If we take the limit as $2a/b \to \infty$ of E_2/E as given by (19a), we see that for a rectangular array (using (20) and letting Area = bd)

$$\lim_{2a/b\to\infty} E_2/E = \frac{1}{1 + 2a/d} \tag{21}$$

Similarly, for a diamond-shaped array with no overlap (i.e., $d > 4a$) where Area = bd/2 , we see that

$$\lim_{2a/b\to\infty} E_2/E = \frac{1}{1 + 4a/d} \tag{22}$$

For a diamond-shaped array with crack overlap, it is observed that B_1^I does not satisfy (20). Rather, it is found that B_1^I is dependent on both b/a and d/a as $b/a \to 0$, and as shown by Fig. 16, for a diamond-shaped array where $d < 4a$,

$$\lim_{b/a\to 0} \frac{B_1^I}{b/a} = \frac{d}{4\pi a} \tag{23}$$

Substituting (23) into (19a) and letting Area = bd/2 , we see that

$$\lim_{2a/b\to\infty} E_2/E = 1/2 \tag{24}$$

as is apparent from Fig. 13.

It is interesting to observe that

$$\lim_{2a/b\to\infty} E_2/E \geq 1/2 \tag{25}$$

for not only the diamond-shaped array with overlap, but also for the rectangular array, where $d > 2a$ in (21), and for the diamond-shaped array without crack overlap, where $d > 4a$ in (22).

It is similarly noted that

$$\mu_{23}/\mu \geq 1/2 \tag{26}$$

for the two doubly-periodic arrays of cracks investigated in this paper.

The change in strain energy per crack for the Mode 2 case may be shown to increase to infinity as the crack spacing tends to zero for both arrays. Hence, as can be seen in Figs. 14 and 15,

$$\lim_{2a/b\to\infty} \mu_{12}/\mu = 0 \tag{27}$$

It would be useful to determine within what range of crack spacings it may be assumed that the crack interaction does not significantly affect the calculation of the effective elastic constants. As an example, let us compare the values of E_2/E for a rectangular array found in this study with the values one would obtain if it were assumed that each crack changed the strain energy by the same amount as an isolated crack (i.e., assume $B_1^I = 1.0$ in (19a)). The results of such a comparison are presented in Fig. 17, where it can be seen that the assumption of no crack interaction is reasonably good only for widely spaced cracks. In general, this value of E_2/E is lower than the value found when crack interaction is accounted for.

5. Summary of Results

It has been shown that a solid containing either a rectangular or diamond-shaped doubly-periodic array of cracks, subjected to an applied in-plane shear stress, is generally weaker (i.e., prone to fracture at lower levels of stress) than a similar body containing a single crack of the same length. That is, the Mode 2 stress intensity factor increases with increasing crack density. Also, the effective shear modulus is reduced considerably by the presence of the cracks, and it tends to zero when the cracks are very closely spaced.

By contrast, if the solid with cracks is subjected to tensile stress normal to the crack faces, the body may be more likely or less likely to fracture than a similar body with a single crack, depending on the shape of the crack pattern. In general, unless the distance separating the rows of cracks is much greater than the distance separating adjacent collinear cracks on the same line, the interaction between cracks in the Mode 1 case is such that the stress intensity factor is reduced, and hence the body is stronger. The reduction of the stress intensity factor is found to be greatly magnified if there is crack overlap in the diamond-shaped array. It was also found that the effective Young's modulus of the material in the direction normal to the crack faces is diminished by the cracks, but it does not fall below one half of the actual value of E.

The results found for the Mode 3 (anti-plane strain) stress intensity factor were analogous to those found for the Mode 1 case, and it was also shown that the shear modulus for anti-plane strain is not less than one half of the actual value, even for very closely spaced cracks.

REFERENCES

1. KOITER,W.T.: An Infinite Row of Collinear Cracks in an Infinite Elastic Sheet, Ingenieur-Archiv., 28, 163-172(1959).

2. PARIS,P.C., SIH,G.C.: Stress Analysis of Cracks, Symposium on Fracture Toughness Testing and Its Applications, STP 381, American Society for Testing Materials, Philadelphia, 30-83(1965).

3. LOUAT,N.: The Distribution of Dislocations in Stacked Linear Arrays, Philosoph. Magazine, 8, No. 91, 1219-1224(July 1963).

4. KOITER,W.T.: An Infinite Row of Parallel Cracks in an Infinite Elastic Sheet, Problems in Continuum Mechanics, English ed., (RADOK,J.E.M., ed.), Society for Industrial and Applied Mathematics, Philadelphia, 246-259(1961).

5. ICHIKAWA,M., OHASHI,M., YOKOBORI, T.: Interaction between Parallel Cracks in an Elastic Solid and Its Effect on Fracture, Reports of the Research Institute for Strength and Fracture of Materials, Tohoku Univ., Sendai, 1, No. 1, 1-14(May 1965).

6. BENTHEM,J.P., KOITER, W.T.: Asymptotic Approximations to Crack Problems, Mechanics of Fracture,Vol. 1: Methods of Analysis and Solutions of Crack Problems (SIH,G.C., ed.),pp.131-178. Leyden: Noordhoff. 1973.

7. BILBY,B.A., ESHELBY,J.D.: Dislocations and Theory of Fracture, Fracture: An Advanced Treatise (LIEBOWITZ,H., ed),Vol. 1,pp. 99-182. New York: Academic Press. 1968.

8. DELAMETER,W.R., HERRMANN,G.: Weakening of an Elastic Solid by a Rectangular Array of Cracks, Stanford University SUDAM Report 73-2.

Acknowledgements

This work was supported in part by AFOSR Grant 70-1905 and in part by the National Science Foundation through the Center for Materials Research at Stanford University.

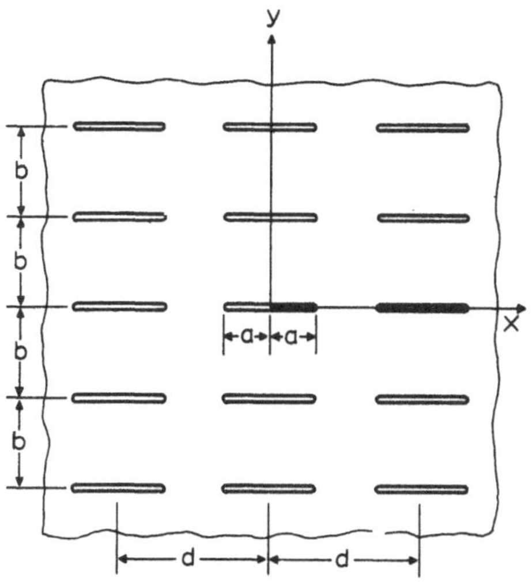

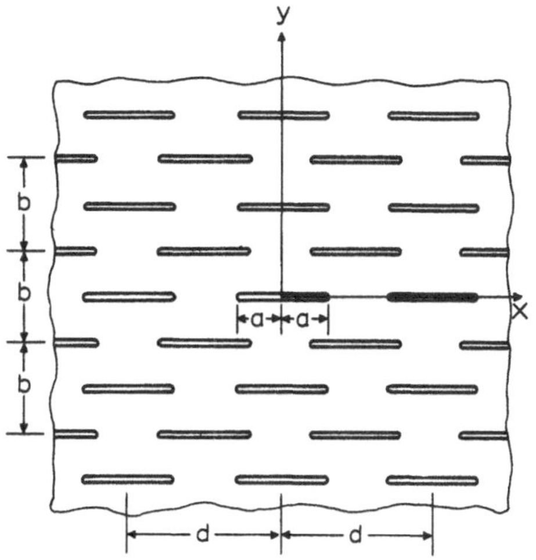

Fig. 1 (a) A rectangular array of cracks (b) a diamond-shaped array of cracks

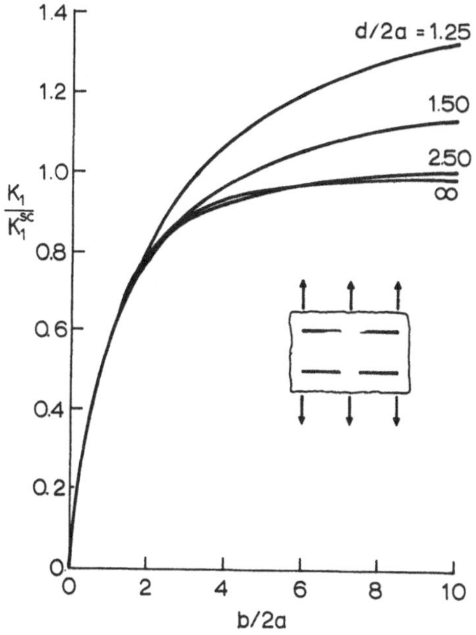

Fig. 2 K_1/K_1^{sc} for a body containing a rectangular array of cracks

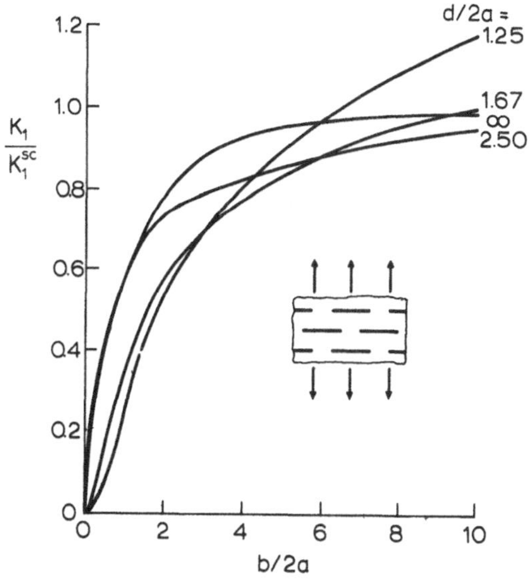

Fig. 3 K_1/K_1^{sc} for a body containing a diamond-shaped array of cracks

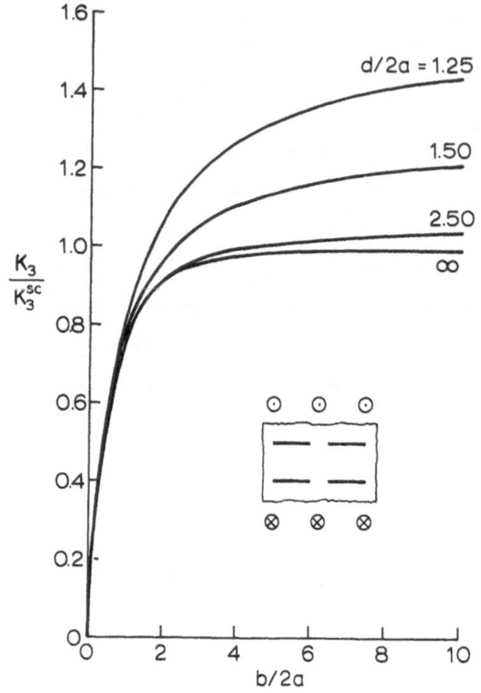

Fig. 4 K_3/K_3^{sc} for a body containing a rectangular array of cracks

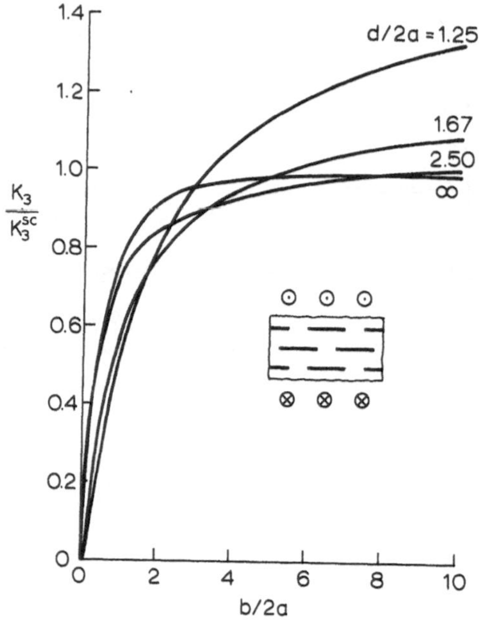

Fig. 5 K_3/K_3^{sc} for a body containing a diamond-shaped array of cracks

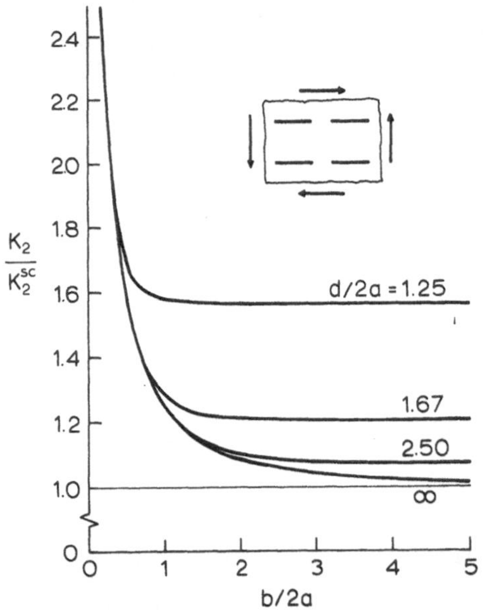

Fig. 6 K_2/K_2^{sc} for a body containing a rectangular array of cracks

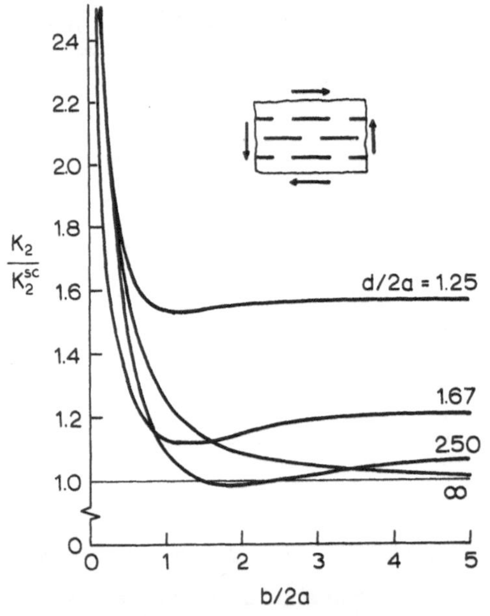

Fig. 7 K_2/K_2^{sc} for a body containing a diamond-shaped array of cracks

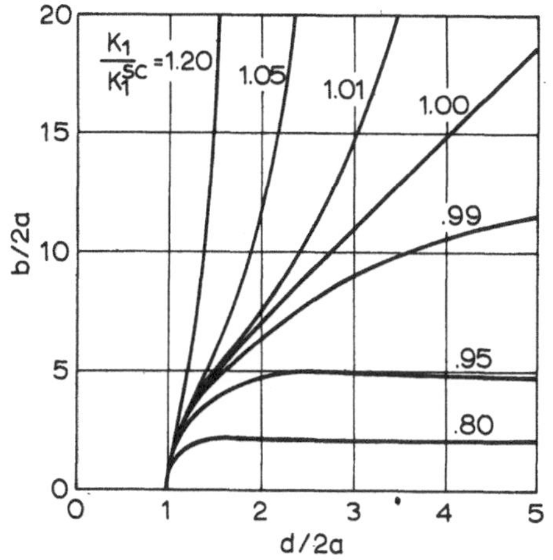

Fig. 8 The relationship between b/2a and d/2a for certain constant values of K_1/K_1^{sc} for a rectangular array of cracks

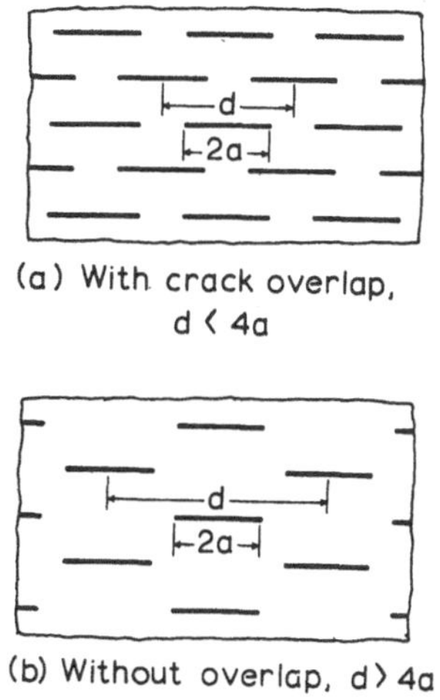

Fig. 9 A diamond-shaped array of cracks, (a) with crack overlap, (b) without crack overlap

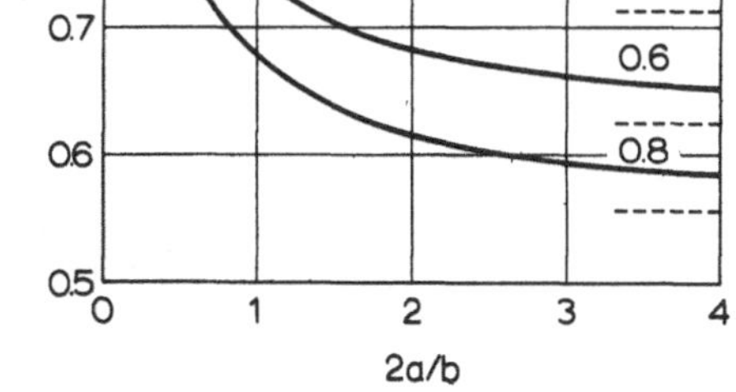

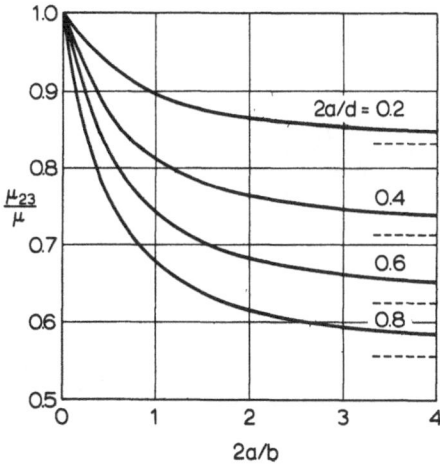

Fig. 10 μ_{23}/μ for an elastic body containing a rectangular array of cracks

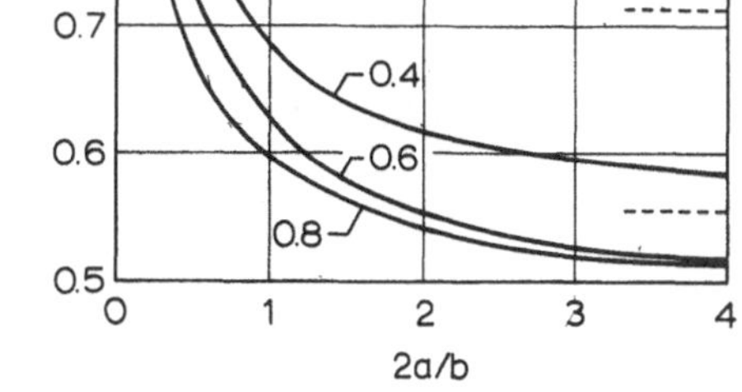

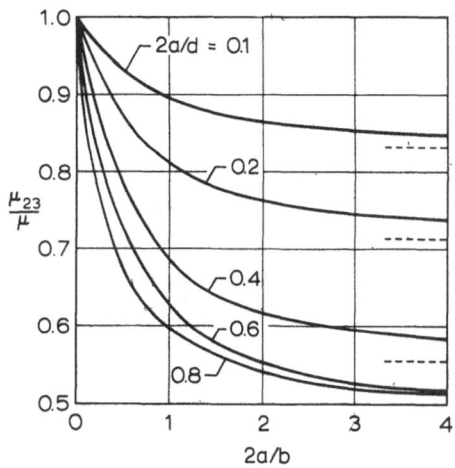

Fig. 11 μ_{23}/μ for an elastic body containing a diamond-shaped array of cracks

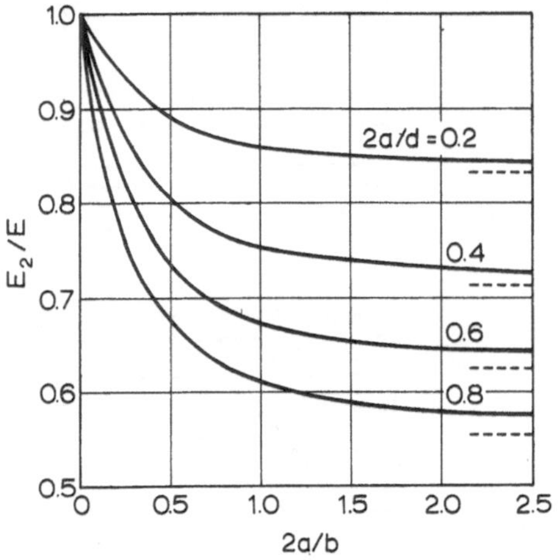

Fig. 12 E_2/E ($= \nu_{21}/\nu$) for an elastic sheet containing a rectangular array of cracks

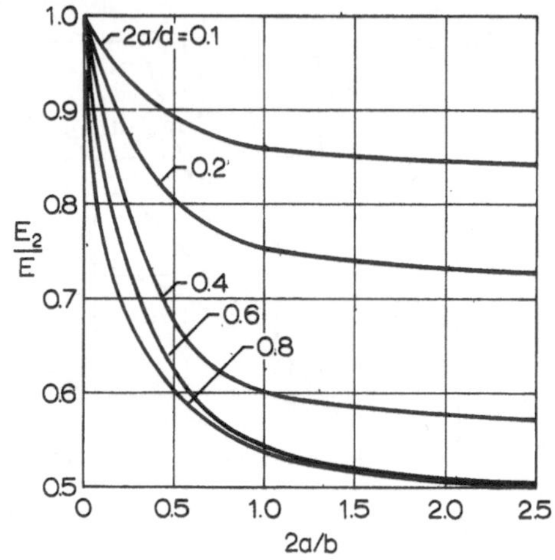

Fig. 13 E_2/E ($= \nu_{21}/\nu$) for an elastic sheet containing a diamond-shaped array of cracks

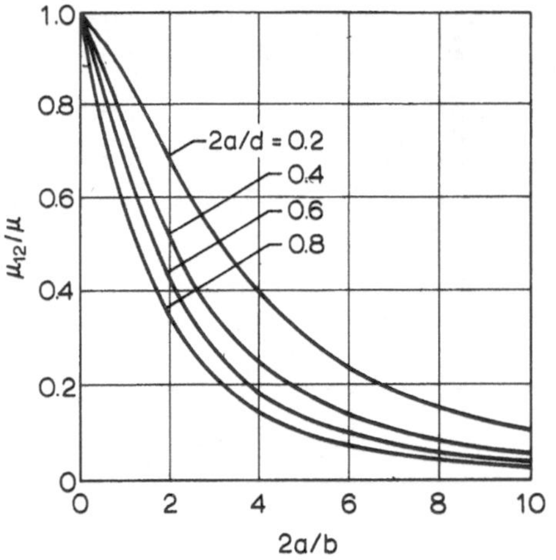

Fig. 14 μ_{12}/μ for an elastic sheet containing a rectangular array of cracks, $\nu = 0.3$

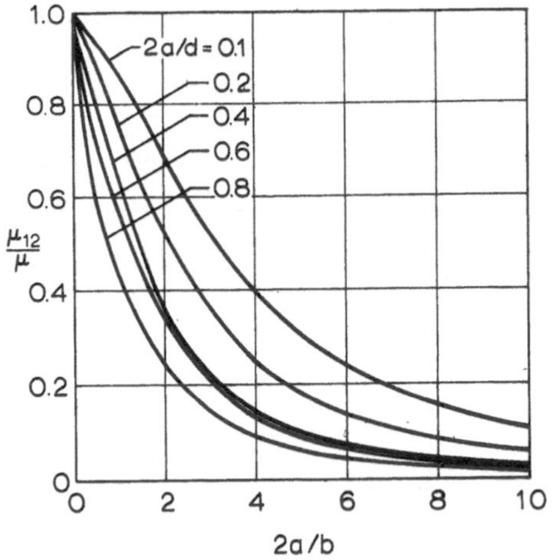

Fig. 15 μ_{12}/μ for an elastic sheet containing a diamond-shaped array of cracks, $\nu = 0.3$

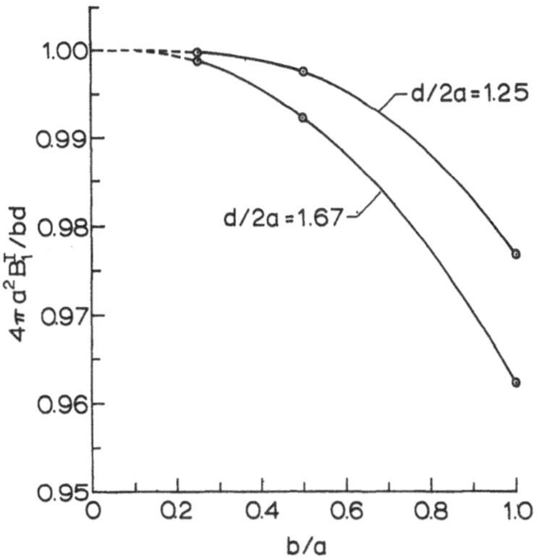

Fig. 16 $4\pi a^2 B_1^I/bd$ vs. b/a for a diamond-shaped array of cracks, illustrating that

$$\lim_{b/a \to 0} \frac{B_1^I}{b/a} = \frac{d}{4\pi a}$$

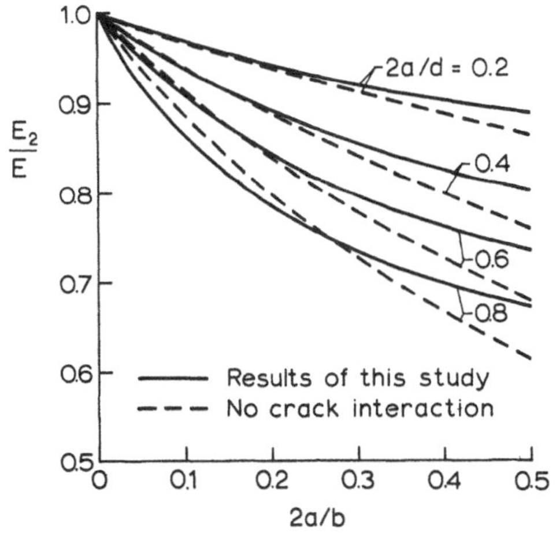

Fig. 17 E_2/E for an elastic sheet containing a rectangular array of cracks comparing results of this study with results obtained if absence of crack interaction is assumed

DYNAMIC THERMAL SHOCK RESISTANCE

H. BARGMANN, Geneva and Vienna

1. Introduction

When DUHAMEL /1/, in 1835, had laid down the foundations of thermo-elasticity, he had derived already the coupled heat conduction equation as well as the corresponding equations of motion. With respect to thermally induced waves and vibrations, however, he asserted the following: "Il est donc permis, surtout à cause de la lenteur avec laquelle s'opère toujours le refroidissement, de négliger complètement ces petits mouvemens des molécules autour de leur position d'équilibre et de considérer l'equilibre comme ayant rigoureusement lieu à chaque instant, et variant avec la propagation intérieure de la chaleur..." and, moreover, he remarked "...que ces mouvemens vibratoires produiraient en chaque point des dilatations et condensations alternatives, dont les effets tendraient à se compenser". Thus the time rate of change of temperature has been considered slow enough so that inertia effects could be disregarded in the equations of thermo-elasticity.

The dynamic problem was studied next by DANILOVSKAYA /2/, in 1950, cf. for example Ref. 3. She solved the famous (uncoupled) problem of the half-space $x \geq 0$ subjected to a sudden step in surface temperature $T(0,t) = T_0U(t)$, resulting in a propagating stress jump $\sigma_x = E\alpha T_0/(1 - 2\nu)$, where α is the coefficient of thermal expansion, and E and ν are Young's modulus and Poisson's ratio, respectively. But when the effect of a finite heating time t_0 on decreasing the magnitude of the jump had been examined by STERNBERG and CHAKRAVORTY /4/ in 1959, DANILOVSKAYA's problem appeared to be of rather academic interest: for carbon steel, for example, for a rise-time $t_0 = 10^{-12}$ sec only, the jump is already reduced to 14% of the above value for $t_0 = 0$. Thus, in similar problems of surface heating of massive bodies, for practical purposes it is sufficient to assume the quasi-static state.

For slender structures under surface heating, however, such as beams, plates, and shells, a possible importance of dynamic effects has been recognized by BOLEY /5/ in 1956: there, particularly with thin plates, gross, thermally induced vibrations may be possible. In order to assess the dynamic effect a simple formula has been proposed from rather general considerations, for the ratio of the maximum dynamic to quasi-static (bending) deflection of heated beams and plates /6/,

$$\frac{w_{dyn}}{w_{st}} = 1 + \frac{1}{\sqrt{1 + (t_T/t_M)^2}} ,$$

where the characteristic mechanical time t_M is usually the lowest natural period, and the characteristic thermal time $t_T = h^2/\kappa_v$ in problems of surface heating, where h is a characteristic length of the body, and κ_v is the thermal diffusivity at constant deformation.

In contrast to surface heating, the rapid, direct, internal heat supply, for example by "instantaneous" dumping of radiation or particles into a structure, is an effective means for inducing waves also in massive bodies, without depending upon thermal conduction /7/. This was certainly not foreseen by DUHAMEL - he had no source term in his heat equation. Today, this mechanism is of importance for both massive and slender bodies; the propagation set up by thermal disturbances is studied. Heat conduction may generally be neglected, the thermal time t_T being much greater than both the mechanical time and the energy deposition time, t_M and t_0, respectively. Problems of this kind have been investigated by a number of authors. The stress analysis of a centrally heated elastic-plastic disk by PARKUS /8/ in 1954 appears to be the first. For further references cf., for example, Ref. 9.

2. The Basic Problem of Radiation Heating

In radiation heating problems, a finite rise-time t_0 of the temperature corresponds to the realistic case of a finite energy deposition time.

It has a significant effect on the maximum stress created in a structure. This may be illustrated for the basic problem of a finite rod of length L, freely suspended, rapidly exposed to a uniform high-temperature field T_0 inside. In other words,

$$\frac{\partial^2\sigma}{\partial\xi^2} - \frac{\partial^2\sigma}{\partial\tau^2} = \frac{\partial^2\Theta}{\partial\tau^2} , \qquad \sigma(-\xi,\tau) = \sigma(\xi,\tau) , \qquad 0 \le \xi \le 1 , \ \tau > 0$$

$$\sigma(\xi,0) = \frac{\partial\sigma}{\partial\tau}(\xi,0) = 0 , \qquad 0 \le \xi \le 1 \tag{1}$$

$$\sigma(1,\tau) = 0 , \qquad \tau > 0$$

where

$$\Theta = T/T_0 , \quad \sigma = \sigma_x/(E\alpha T_0) , \quad \xi = 2x/L , \quad \tau = 2ct/L , \quad c^2 = E/\rho , \tag{2}$$

with temperature T and longitudinal stress σ_x; and where c is known as the bar velocity.

For the time-ramp-type temperature field

$$\Theta(\tau) = \begin{cases} 0 & \text{if } \tau \le 0 \\ \tau/\tau_0 & \text{if } 0 \le \tau \le \tau_0 \\ 1 & \text{if } \tau_0 \le \tau \end{cases} \tag{3}$$

with rise-time τ_0, the solution of Eqs. (1) can be given in closed form /10/

$$\sigma(\xi,\tau;\tau_0) = \frac{1}{\tau_0}\left\{ -\tau + \int_{\min[0,\tau-(1-\xi)]}^{\tau-(1-\xi)} f(u)\,du + \int_{\min[0,\tau-(1+\xi)]}^{\tau-(1+\xi)} f(u)\,du \right\} \qquad \text{if } 0 < \tau < \tau_0$$

$$\sigma(\xi,\tau;\tau_0) = -1 + \frac{1}{\tau_0}\left\{ \int_{\max\{\tau-(1-\xi)-\tau_0,\ \tau-(1-\xi)-\max[0,\tau-(1-\xi)]\}}^{\tau-(1-\xi)} f(u)\,du + \int_{\max\{\tau-(1+\xi)-\tau_0,\ \tau-(1+\xi)-\max[0,\tau-(1+\xi)]\}}^{\tau-(1+\xi)} f(u)\,du \right\} \qquad \text{if } \tau_0 < \tau \tag{4}$$

where

$$\int_0^\tau f(u)\,du = f_1(\tau) + \tau/2 \tag{5}$$

with

$$f_1(\tau) \equiv \begin{cases} \tau - 2n & \text{if} \quad 2n < \tau < 2n + 1 \\ 2(n + 1) - \tau & \text{if } 2n + 1 < \tau < 2(n + 1) \end{cases} \tag{6}$$

when $n = 0, 1, 2, \ldots$.

Thus, owing to the free thermal expansion at the ends, unloading waves are travelling from the ends into the compressed rod. In particular, for very slow temperature increase, $\tau_0 \to \infty$, the expression for the longitudinal stress σ turns into the quasi-static solution $\sigma \equiv 0$.

The maximum (compressive and tensile) stresses created are

$$\sigma_{max} = \begin{cases} \pm 1 & \text{if } \tau_0 \leq 2 \\ \pm 2/\tau_0 & \text{if } \tau_0 \geq 2 \end{cases} \tag{7}$$

or

$$\frac{|\sigma_x|_{max}}{E\alpha T_0} = \begin{cases} 1 & \text{if } t_0 \leq t_M \\ t_M/t_0 & \text{if } t_0 \geq t_M \end{cases} \tag{8}$$

where the maximum temperature T_0 is reached at time t_0 and then maintained, according to a time-rectangular heat supply; and the mechanical time is $t_M = L/c$. Thus for an extremely fast temperature rise, $t_0 < t_M$, maximum stresses have the constant, absolute value of a compressive stress in a rod fixed at the ends, independently of length and rise-time. For slower, but still fast temperature rises, the maximum stresses are proportional to the length L, and inversely proportional to the rise-time t_0.

Experimental results in this field of radiation heating are reported by several authors (cf. Ref. 9 for further references). A striking illustration has been presented by AVERY, KEEFE, BREKKE, and FINNIE /11/, and is shown in Fig. 1.

Fig.1. High speed photography, at time t = 2 msec, records pieces of granite being exploded away from a 1 cm slab (dark vertical bar running across the photograph) which has been bombarded with a single intense burst of electrons with rise-time t_0 = 50 nsec. The electron burst created stress waves in the granite which not only disintegrated the right side of the rock where the electrons entered but also travelled through the slab and fractured the left side as well. (Photo from AVERY, KEEFE, BREKKE, and FINNIE: Shattering rock with intense bursts of energetic electrons /11/, by author's courtesy.)

3. Thermal Shock Resistance of Structures

From the exact result, Eq. (8), there follows a simple and natural measure for dynamic thermal shock resistance of structures under radiation heating,

$$R_r = \frac{c_V \sigma_{adm}/\rho}{\alpha\sqrt{E/\rho}} , \tag{9}$$

which is proportional to the specific heat c_V and to the specific strength, inversely proportional to the coefficient of thermal expansion and to the square root of the specific stiffness. Since from the heat equation, in the present case, we have $T_0/t_0 = r_0/c_V$, there follows

$$R_r \equiv L r_0 . \tag{10}$$

Thus, Eq. (9), for a given length L of the rod, measures the admissible heat supply r_0 (per unit mass and unit time); for a prescribed heat supply it gives the admissible length of the rod, so that the stress σ_x will not exceed the admissible (tensile or compressive) strength value σ_{adm}.

The preceding equations have been derived under the assumption that $t_0, t_M \ll t_T = h^2/\kappa_V$, h being a characteristic radius of the cross-section of the rod. Typical values of situations around high-energy particle accelerators are $t_0 \sim 10^{-5}$ sec, $t_M \sim 10^{-5}$ sec, $t_T \sim 10^{-1}$ sec /10/. Again it should be stressed that the equations are exact whenever the assumption of a homogeneous time-ramp-type temperature field (corresponding to a time-rectangular heat supply) are fulfilled. Since it is the heat supply per unit mass r_0, as it has been assumed, which is more likely to be prescribed in radiation problems, rather than the heat supply per unit volume ρr_0, Eq. (9) provides a natural criterion for dynamic thermal shock resistance.

4. Representative Materials

A number of materials are compared with respect to their dynamic thermal shock resistance under radiation heating R_r. The frequently used quasi-static thermal shock parameter

$$R_s = \frac{K\sigma_{adm}}{\alpha E} \tag{11}$$

with conductivity K, measures, in problems of surface heating, an admissible heat flux. It is evident from Table 1 that, in general, from R_s nothing can be concluded about the thermal shock resistance of a structure under dynamic conditions. Then the dynamic parameter R_r provides a natural and more rational measure than formulae of the type of Eq. (11), or the even vaguer and obscure criterion /12/ which still measures thermal shock resistance in terms of "excellent", "good", "fair", or "poor".

Table 1

Static and dynamic thermal shock resistance of several materials, after Eqs. (11) and (9), respectively (for the particular heat supply r_0 = 12 cal g^{-1} μsec^{-1})

		Thermal shock resistance	
		dynamic R_r (cal g^{-1} μsec^{-1} cm)	static R_s (kcal h^{-1} cm^{-1})
Carbon fibre	C	1550	2530
Beryllium	Be	115	4
Pyrolitic graphite	C	57	1610
Zirconia	ZrO_2	44	3
Graphite	C	35	160
Beryllia	BeO	14	6
Alumina	Al_2O_3	8	5
Stainless steel	18Cr8Ni	5	12
Tungsten	W	4	385

References

1. DUHAMEL, J.-M.-C.: Second mémoire sur les phénomènes thermomécaniques. Journal de l'Ecole polytechnique, Paris, 25, 31 (1837).
2. DANILOVSKAYA, V.Y.: Temperature stresses in an elastic semi-space due to a sudden heating of its boundary (in Russian). Prikl. Mat. Mekh. (1950).
3. PARKUS, H.: Instationäre Wärmespannungen, Wien: Springer. 1959.
4. STERNBERG, E., and J.G. CHAKRAVORTY: On inertia effects in a transient thermoelastic problem. J. Appl. Mech. 26 (1959).
5. BOLEY, B.A.: Thermally induced vibrations of beams. J. Aeronaut. Sci. 23, 179-181 (1956).
6. BOLEY, B.A.: Approximate analyses of thermally induced vibrations of beams and plates. J. Appl. Mech. 39, 212-216 (1972).
7. LYONS, W.C.: Comments on heat induced vibrations of elastic beams, plates, and shells. AIAA J. 4, 1502-1503 (1969).
8. PARKUS, H.: Stress in a centrally heated disc. Proc. 2nd U.S. Nat. Congr. Appl. Mech. p. 307. 1954.

9. BARGMANN, H.: Recent developments in the field of thermally induced waves and vibrations. 2nd Int. Conf. Struct. Mech. Reactor Tech. Sept. 10-14, Berlin (1973). Nuclear Engng. Design (in press).
10. BARGMANN, H.: Stress waves in elastic rods induced by radiation heating. Nuclear Engng. Design (in press).
11. AVERY, R.T., D. KEEFE, T.L. BREKKE, and I. FINNIE: Shattering rock with intense bursts of energetic electrons. In Proc. Particle Accelerator Conf. March 5-7, San Francisco, 1973.
12. HAUCK, J.E., (ed.): Materials selector 73. vol. 76, 4. Stamford, Conn.: Reinhold Publ. Co. Inc. 1972.

PLASTOKINETICS OF METAL FORMING
(Review)

H.LIPPMANN, Karlsruhe

1. Introduction

While the term "dynamic plasticity" occasionally refers also to the rate dependence of the uniaxial yield stress Y, or of the yield stress in shear K we shall focus our attention mainly to the proper inertial terms

$$\rho\dot{v}^j = \rho\left(\frac{\partial v^j}{\partial t} + v^l \partial_l v^j\right) \tag{1}$$

where ρ is the material density, $v^j = \dot{x}^j$ denote the velocities with respect to any contra-variant co-ordinates x^j, t the time, $\dot{} = \frac{d}{dt}$ the total time differentiation, and ∂_l the co-variant derivative with respect to x^l. In general considerations of this sort, RICCI's summation convention will be used.

To express the inertial terms (1) in a standardized manner let us introduce a standard length H, a standard velocity c, a standard acceleration $\dot{c}$, as well as an average shear strength $\bar{K}$ of the considered problem all of them being constant in space, and define the related velocities w^j, accelerations a^j, as well as coordinates ξ^j according to

$$w^j = \frac{v^j}{c}\,, \quad a^j = \frac{1}{\dot{c}}\frac{\partial v^j}{\partial t}\,, \quad \xi^j = \frac{x^j}{H} \tag{2}$$

respectively. If then $\bar{\partial}_j$ denotes the co-variant derivative with respect to ξ^j we obtain from equ.(1) that

$$\frac{\rho H}{2\bar{K}} \overset{+}{v}^j = \beta a^j + \varepsilon w^1 \bar{\partial}_1 w^j \tag{3}$$

where the dimensionless quantities /2,12/

$$\varepsilon = \frac{\rho c^2}{2\bar{K}} , \quad \beta = \frac{\rho H \dot{c}}{2\bar{K}} \tag{4}$$

must at least have the order of magnitude 1 for inertial terms to become influential. This happens in the domain of technically important metal forming processes, with high velocity compression (forging) and high rate of energy (explosive) sheet metal forming so that the subsequent analysis will deal with them. Dynamics in cutting has to do with elasticity and vibrations rather than with plasticity though concentrated plastic shear zones are well combined with an accelerated flow /2/. Its effect remains, however, mostly local. As well, the large field of dynamic plasticity in application to material testing, civil or structural engineering, and similar problems shall be excluded here; we refer instead to two text-books /3,4/ or to original literature.

The constitutive equations are to be formulated as usual in plasticity /1/ using any generalized stresses Q_j (which could re-present tensorial stresses $\sigma^j_{\cdot k}$, or bending moments in plate theory, or similar terms ordered as a linear array), and associated rates q^j (like strain rates $\lambda^{\cdot k}_{j}$, rates of curvature, etc.) so that

$$\Lambda = Q_j q^j \geqq 0 \tag{5}$$

expresses the rate of dissipation per unit volume. In case of plastic flow, a "yield condition"

$$f(Q,K) = 0 \ , \quad \text{where} \quad f(0,K) \leqq 0 \tag{6}$$

must be fulfilled by the stresses $Q = (Q_j)$ in which f defines a convex hyper-surface in the space of the Q_j's. Then the "yield rule" (v.MISES potential rule, normality rule) becomes

$$q^j = \kappa \frac{\partial f}{\partial Q_j} \ ; \quad \kappa \geqq 0 \tag{7}$$

where the positivity of the new variable κ, and of the rate of dissipation Λ, equ.(5), are equivalent. At points Q where f is not differentiable, equ.(7) may be generalized according to KOITER /1/.

States of stress for which

$$f(Q,K) \leqq 0 \tag{8}$$

are called "admissible"; they belong in case $f < 0$ to elastic parts of the body, or to rigid ones if elasticity would be disregarded ("rigid-plastic" material).

2. Some Rigorous Solutions to Kinetically Stationary Problems

2.1 General

Problems in which with regard to time, the acceleration parameter β is not independent but a function of ε alone,

$$\beta = \beta(\varepsilon) \ , \tag{9}$$

will be called "kinetically stationary". Despite this considerable simplification there is no three-dimensional complete, and realistic solution known exept for pure shear (HAGEN-POISEUILLE flow) which will be discussed later under non-stationary conditions even. Referring to metal forming, NAJAR's polar-like flow /5/ could be an approach to wire drawing but is not since it does not allow for the transition between the plastic region and the wire at the entrance and the exit of the die. As an elementary analysis shows that inertia may be neglected without that /2/ we shall subsequently consider two different applications which, despite their shortcomings refer at least to proper kinetic situations.

2.2 Plane-Strain Compression of a Layer

The layer as shown in Fig.1 is compressed between two rigid parallel plates $y = \pm H$ so that

$$c = -\dot{H} = \text{const} \ ; \quad \dot{c} = 0 \ . \tag{10}$$

We assume the material to be homogeneous, isotropic, and "ideally plastic"

$$Y = \text{const} \quad \text{or} \quad K = \bar{K} = \text{const}, \tag{11}$$

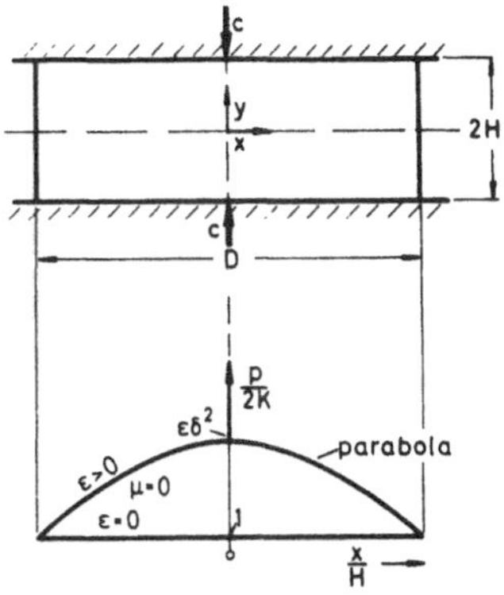

Fig.1. Plane-strain compression; kinetic pressure distribution at the ram (NAJAR)

respectively, and consider the upper right quarter plane $x \geqq 0$, $y \geqq 0$ alone. Then we elect the generalized forces Q_j, and rates q^j in accordance with the physical stresses, and strain-rates respectively,

$$\left.\begin{aligned} Q_1 &= \sigma_x, \quad Q_2 = \sigma_y, \quad Q_3 = 2\sigma_{xy} = 2\tau \\ q^1 &= \lambda_x, \quad q^2 = \lambda_y, \quad q^3 = \lambda_{xy} = \chi \end{aligned}\right\} \tag{12}$$

so that equ.(5) holds. For an incompressible material the yield condition f must not contain the mean stress and thus depend on the radius of MOHR's circle only,

$$f = (\sigma_x - \sigma_y)^2 + (2\tau)^2 - (2K)^2 = 0 . \tag{13a}$$

The yield rule (7) reduces after elimination of κ to

$$\lambda_x + \lambda_y = 0 \quad \text{(incompressibility)}, \tag{13b}$$

$$\frac{\sigma_x - \sigma_y}{2\tau} = \frac{\lambda_x - \lambda_y}{2\chi} \tag{13c}$$

(coincidence of principal axes) while using the cartesian velocities $u = v_x$, $v = v_y$ we obtain

$$\lambda_x = \frac{\partial u}{\partial x}, \quad \lambda_y = \frac{\partial v}{\partial y}, \quad \chi = \frac{1}{2}\left(\frac{\partial u}{\partial y} + \frac{\partial v}{\partial x}\right). \tag{13d}$$

Equilibrium becomes using (1)

$$\frac{\partial \sigma_x}{\partial x} + \frac{\partial \tau}{\partial y} = \rho\left(\frac{\partial u}{\partial t} + \frac{\partial u}{\partial x}\, u + \frac{\partial u}{\partial y}\, v\right) , \tag{13e}$$

$$\frac{\partial \tau}{\partial x} + \frac{\partial \sigma_y}{\partial y} = \rho\left(\frac{\partial v}{\partial t} + \frac{\partial v}{\partial x}\, u + \frac{\partial v}{\partial y}\, v\right) . \tag{13f}$$

Equs.(13a-f) constitute the governing relations of the problem. Already in 1923, PRANDTL gave a solution for static flow $\rho = 0$ assuming that τ does not depend on x, and that $|\tau|$ reaches its maximum value K at the platens $y = \pm H$. It was generalized to kinetics by BYKOVCEK /6/, and subsequently to more general (though still simplified) frictional conditions as defined by a constant coefficient μ; $0 \leqq \mu \leqq 1$ according to

$$\tau = -\mu K \quad \text{if} \quad u > 0, \quad \text{at} \quad y = 1 \tag{14}$$

by NAJAR /7/. A result which is to fulfil as well

$$\left.\begin{aligned} \tau = 0\ , \quad v = 0 \quad \text{at} \quad y = 0\ , \\ v = -c \quad \text{at} \quad y = H \end{aligned}\right\} \tag{15}$$

can by substitution into equs.(12-15),(5), be seen regarding (4) to be

$$u = c\left\{\frac{x}{H} + \frac{1}{\varepsilon}\left[g'\left(\frac{y}{H}\right) + A\right]\right\} , \tag{16a}$$

$$v = -c\,\frac{y}{H} , \tag{16b}$$

$$\sigma_x = 2K\left\{\varepsilon\left(\frac{x}{H}\right)^2 + A\,\frac{x}{H} - B + \sqrt{1-4g^2\left(\frac{y}{H}\right)}\right\}, \tag{16c}$$

$$\sigma_y = 2K\left\{\varepsilon\left(\frac{x}{H}\right)^2 + A\,\frac{x}{H} - B\right\} , \tag{16d}$$

$$\tau = 2K\, g\left(\frac{y}{H}\right) , \tag{16e}$$

where A,B are time-independent integration constants, and the function $g(\eta)$ has to obey the ordinary second order differential equation

$$g''(\eta) = \frac{4g(\eta)\varepsilon}{\left|\sqrt{1-4g^2(\eta)}\right|} \tag{17a}$$

as well as the boundary conditions following from equs.(14,15),

$$g(1) = -\frac{\mu}{2}\ ; \quad g(0) = 0\ . \tag{17b}$$

Though there is a unique integral g of equ.(17a) which even fulfills as necessary, always the inequality $|g| \leqq 1$ it can in a rather complicated manner be expressed in terms of elliptic functions only. We will not elaborate it but mention that

$$\left.\begin{aligned} g(\eta) &= -\frac{\mu}{2}\eta \quad \text{if} \quad \varepsilon = 0\ , \\ g(\eta) &\equiv 0 \qquad \text{if} \quad \mu = 0\ . \end{aligned}\right\} \tag{18}$$

Unfortunately, our problem requires further boundary conditions to be regarded too,

$$\left.\begin{aligned} u = 0 \quad &\text{if} \quad x = 0\ , \\ \sigma_x = \tau = 0 \quad &\text{if} \quad x = \frac{D}{2} \end{aligned}\right\} \tag{19}$$

which is impossible if $\varepsilon \neq 0$ or $\mu \neq 0$. Thus following an often applied procedure one only claims that they hold in average,

$$\left.\begin{aligned} \int_{-H}^{H} u\,dy = 0 \quad &\text{if} \quad x = 0\ , \\ \int_{-H}^{H} \sigma_x dy = \int_{-H}^{H} \tau\,dy = 0 \quad &\text{if} \quad x = \frac{D}{2}\ . \end{aligned}\right\} \tag{20}$$

Because of symmetry, it will do to fulfil the first two to give

$$A = \frac{\mu}{2}\ , \quad B = \varepsilon\delta^2 + \frac{\mu}{2}\delta + \int_0^1 \sqrt{1 - 4g^2(\eta)}\,d\eta\ , \tag{21}$$

where

$$\delta = \frac{D}{2H}\ . \tag{22}$$

Then it is automatically guaranted that the above solution (16) converges to the static PRANDTL one if $\varepsilon \to 0$, and a rough approximation can be found by assuming (18) to hold also under conditions of kinetics. Equ.(16d) determines the forging pressure $p = -(\sigma_y)_{y=H}$ (cp. Fig.1), and an integration over the upper face of the specimen gives the forging force per unit width,

$$P = 2KD\left\{\frac{2}{3}\varepsilon\delta^2 + \frac{\mu}{4}\delta + \int_0^1 \sqrt{1 - 4g^2(\eta)}\,d\eta\right\}\ . \tag{23}$$

2.3 Axially Symmetric Bulging of a Membrane

The isotropic membrane built in at r = a as shown in Fig.2, is assumed to have been at the beginning t = 0, a flat circular plate with the constant thickness D_o which obtains by a sudden impact (explosion pressure transferred by water) an initial vertical velocity distribution,

$$v = -c\, Z\left(\frac{r}{a}\right), \quad Z(0) = 1\,, \quad Z(1) = 0\,, \tag{24}$$

and deforms free afterwards under the action of its own inertial forces only. The impact as such is not considered so that acceleration need not be prescribed, and ε becomes the only relevant kinetic parameter. The instantaneous shape z=z(r,t) given in polar coordinates r,ψ,z will be combined with a variable thicknes D(r,t).

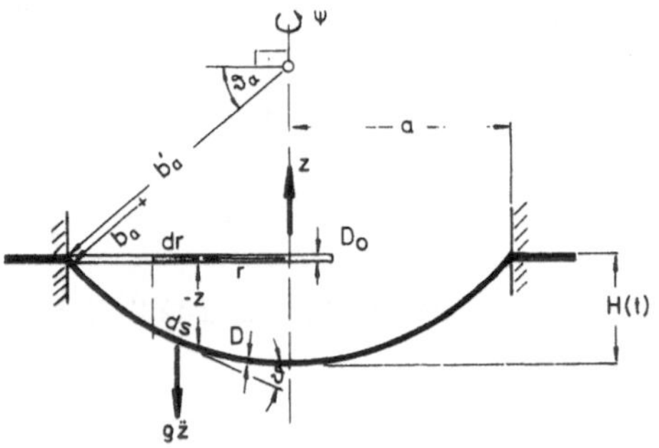

Fig.2. Inertial Bulging

The principal strain rates as defined physically in azimuthal, circumferential, and thickness directions respectively, become

$$\lambda_s = \frac{(ds)^{\cdot}}{ds}\,, \quad \lambda_\psi = \frac{\dot r}{r}\,, \quad \lambda_D = \frac{\dot D}{D} \tag{25}$$

where s denotes the arc length outward. By virtue of the membrane approach, the related physical principal stresses σ_s, σ_ψ,σ_D fulfil

$$\sigma_D \equiv 0\,, \tag{26}$$

as well as the equilibrium conditions (cp. PARKUS /9/),

$$\left.\begin{aligned} \rho\,\ddot{z}\sin\theta - \frac{\sigma_s}{b} - \frac{\sigma_\psi}{b'} &= 0\,, \\ -\rho\,\ddot{z}\cos\theta + \frac{1}{D}\,\frac{\partial(D\sigma_s)}{\partial s} + \frac{\sin\theta}{r}\,(\sigma_s-\sigma_\psi) &= 0\,, \end{aligned}\right\} \tag{27}$$

where b,b' are the principal radii of curvature with respect to meridional, and azimuthal sections respectively (cp. b_a, b_a' at $r = a$, Fig.2). They obey together with the angle of the tangent θ, the geometrical relations

$$\left.\begin{aligned} \frac{1}{b} = \frac{d\theta}{ds}\,, \quad b' = \frac{r}{\cos\theta}\,, \quad \frac{dr}{ds} = \sin\theta\,, \\ \frac{dz}{ds} = \cos\theta\,, \quad \frac{dz}{dr} = \cot\theta\,. \end{aligned}\right\} \tag{28}$$

Following ref./8/ we assume that TRESCA's yield condition holds which means that the maximum shear stress at each point must in case of yielding, assume its critical value K, i.e. (cp./2/)

$$f = \sigma_1 - \sigma_3 - 2K = 0 \quad \text{if} \quad \sigma_1 \geqq \sigma_2 \geqq \sigma_3\;; \tag{29}$$

σ_j being the suitably ordered principal stresses. For the points of singularity $\sigma_1 = \sigma_2$, $\sigma_2 = \sigma_3$ excluded we obtain the yield rule (7) as follows:

$$\lambda_1 = \kappa \geqq 0\,, \quad \lambda_2 = 0\,, \quad \lambda_3 = -\kappa \leqq 0 \quad \text{if} \quad \sigma_1 > \sigma_2 > \sigma_3\,, \tag{30}$$

λ_j denoting the strain-rates associated with the stresses σ_j. It can be shown /8/ that our problem possesses in the case of large but not excessive deformations of a rate-independent material, an integral just of that kind namely, regarding equs. (26) and (29),

$$2K = \sigma_s > \sigma_\psi > \sigma_D = 0\,. \tag{31}$$

This means because of equs.(30),(25) that r = const: Particles move straight downward, and we obtain using $dr = ds\sin\theta$,

$$D = D_o \sin\theta\,. \tag{32}$$

Unfortunately, the agreement of equ.(32) with experimental measurements in explosive forming /10/ is bad probably because of the simplified membrane approach in which bending moments are disregarded, or since water pressure does not only act during the first impact.

Now from equs.(27),(28),(31),(32) we obtain one single partial differential equation to determine the contour $z = z(r,t)$,

$$\frac{\partial}{\partial r}(r K \sin 2\theta) = \rho r \frac{\partial^2 z}{\partial t^2} \tag{33}$$

which could be integrated numerically. Confining our analysis to small or medium size deformations $\theta \approx \frac{\pi}{2}$, $\sin 2\theta \approx 2\partial z/\partial r$ and to ideally plastic material K = const, equ.(33) becomes simply

$$\frac{\partial}{\partial r}\left(r \frac{\partial z}{\partial r}\right) = \frac{\rho}{2K} r \frac{\partial^2 z}{\partial t^2}$$

which linear relation may be solved using equ.(24) by means of

$$z(r,t) = -Z\left(\frac{r}{a}\right) H(t)$$

where

$$H(t) = \frac{2T}{\pi} c \sin\frac{\pi}{2}\frac{t}{T}, \quad Z(\xi) = J_o(\gamma\xi)$$

(J_o: BESSEL function of order 0), and

$$\gamma = 2.405\ldots, \quad T = 0.653\ldots \frac{a}{c}\sqrt{\varepsilon}.$$

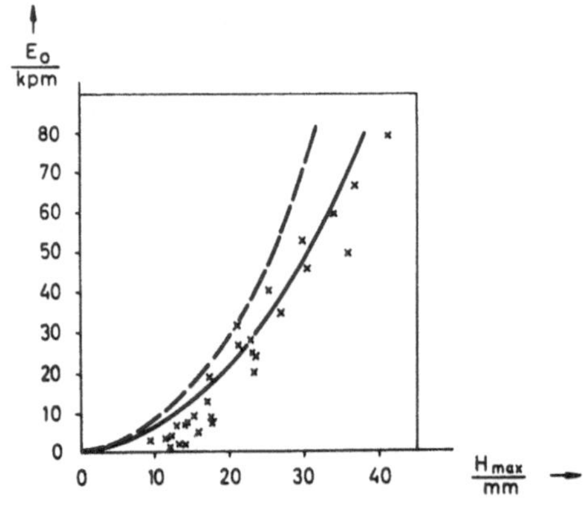

Fig.3. Explosive forming of mild steel.
ρ=803 kp s²/m⁴, a=76,2 mm, D_0=0.9144 mm.
× × × experiments (JOHNSON at al.)
—— theory using K=12.2 kp/mm²(initial value)
---- theory using K=17.8 kp/mm²(mean value $\bar{K}$)

$\varepsilon = \rho c^2/2K$ agrees with equ.(4), while T represents the total duration of the process. Now using equ.(24) the amount of the initial kinetic energy E_o can be evaluated,

$$E_o = 0.423\ldots \rho\, D_o\, a^2 c^2 ,$$

to show satisfactory coincidence with experiments /10/, Fig.3.

3. Perturbation Method

3.1 General, Indentation of the Half Space

The only systematic approach to plasto-kinetics according the author's knowledge, seems to be that of perturbation theory. SPENCER's procedure /11/ for plane deformation of ideally plastic bodies, will be presented here under a little more general aspects.

Making again use of the parameters $\rho, c, H, \bar{K}, \varepsilon, \beta$ which already appeared in equs.(3),(4), and introducing characteristic time constants

$$\overset{0}{T} = H\sqrt{\frac{\rho}{2\bar{K}}} , \quad \overset{1}{T} = \frac{H}{c} = \frac{\overset{0}{T}}{\sqrt{\varepsilon}} , \quad \overset{2}{T} = \sqrt{\frac{H}{\bar{c}}} = \frac{\overset{0}{T}}{\sqrt{\beta}} \tag{34}$$

we form additionally to equ.(2) the standardized strain rates $\gamma_j^{\cdot k}$, stresses $s^j_{\cdot k}$, time $\bar{t}$, and shear strength N respectively according to

$$s^j_{\cdot k} = \frac{\sigma^j_{\cdot k}}{2\bar{K}} , \quad \gamma_j^{\cdot k} = T\lambda_j^{\cdot k} , \quad \bar{t} = \frac{t}{T} , \quad L = \frac{K}{\bar{K}} \tag{35}$$

where T denotes suitably chosen representatives of the above $\overset{i}{T}$. Writing furthermore $f(s^j_{\cdot k}, L) = f(s_k^{\cdot j}, L)$ [1] instead of f(Q,K) but preserving κ as a factor of proportionality we obtain the conditions of equilibrium, compatibility as well as the yield criterion (6) and the yield rule (7) regarding equ. (3), in the following form:

---

[1] Symmetry because of the symmetry $s^j_{\cdot k} = s_k^{\cdot j}$ of stress tensor

$$\left.\begin{aligned} \bar{\partial}^j \, s_j^{\cdot k} &= \beta\, a^k + \varepsilon\, w^l\, \bar{\partial}_l\, w^k \,, \\ \gamma^{j}_{\cdot k} &= \tfrac{1}{2}\left\{ \bar{\partial}_k\, w^j + \bar{\partial}^j\, w_k \right\} , \\ f(s^{j}_{\cdot k}, L) &= 0 \,, \\ \gamma^{j}_{\cdot k} &= \kappa \, \frac{\partial f}{\partial s_j^{\cdot k}} \,, \qquad \kappa \geqq 0 \,. \end{aligned}\right\} \qquad (36)$$

Now the idea is to develop all the relevant quantities $X = s_j^{\cdot k}, a^k, w^l, \gamma^{j}_{\cdot k}$, and as a consequence also f in a power series with respect to β, ε according to

$$X = \underset{pq}{X}\, \varepsilon^p \beta^q \;; \qquad p,q = 0,1,\ldots,\infty, \qquad (37)$$

and to equate in the relations (36), the coefficients of equal powers. For stationary processes $\beta = \beta(\varepsilon)$ all quantities should be developed with respect to ε alone,

$$X = \underset{p}{X}\, \varepsilon^p \qquad (38a)$$

after according to equs.(34),(35),(2) the substitutions

$$\bar{t} = \frac{t}{\overset{1}{T}} \,, \quad \beta\, a^k = \varepsilon\, \frac{\partial w^k}{\partial \theta} \,, \quad \gamma_j^{\cdot k} = \overset{1}{T}\, \lambda_j^{\cdot k} \qquad (38b)$$

have been made. SPENCER /11/ considered instead the case $\varepsilon = \varepsilon(\beta)$, $\beta \neq 0$, in which only β might play a role, and therefore substituted specifically

$$\left.\begin{aligned} &\bar{t} = \frac{t}{\overset{2}{T}} \,, \quad c = \frac{H}{\overset{0}{T}} \quad \text{so that} \quad \varepsilon \equiv 1, \\ &\beta a^k = \sqrt{\beta}\, \frac{\partial w^k}{\partial \theta} \,, \quad \gamma_j^{\cdot k} = \overset{0}{T}\, \lambda_j^{\cdot k} \,. \end{aligned}\right\} \qquad (39a)$$

As a consequence, the approach

$$X = \underset{q}{X}\, \beta^q \qquad (39b)$$

makes only sense if $X = s_j^{\cdot k}$, or f while in case $X' = \gamma_j^{\cdot k}$, w^k, κ one has to assume that

$$X' = \sqrt{\beta}\left(\underset{q}{X'}\,\beta^q\right). \tag{39c}$$

Under all the above circumstances, equs.(36) are split off in a recursive sequence of systems the zero one (belonging to zero powers) is essentially identical with the static one which in this way becomes the starting point for the perturbation method, while the higher ones are linear. Convergence to the true result has never been demonstrated, it fails certainly to hold in the event of impact when β tends to infinity, or if singular points of the yield criterion f should be involved.

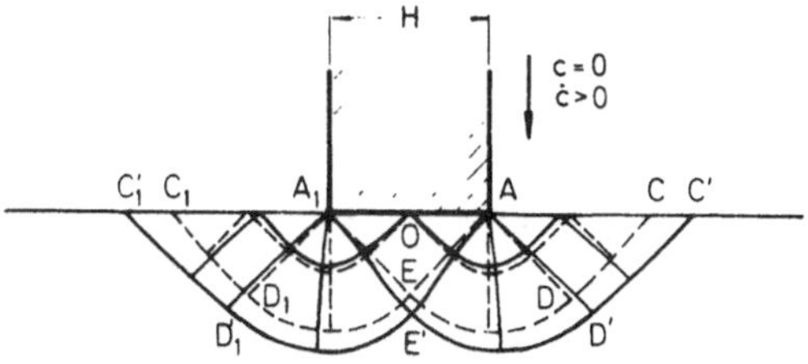

Fig.4. Slip line fields for plane strain indentation(SPENCER)
--- initial (static)
—— linearly disturbed (kinetic: ε=1, β=1/4)

Fig.4 shows the disturbed slip line field (curves inclined by π/4 to the principal stress directions which indicate maximum shear stress, and strain rate) in the plane strain indentation of an isotropic rigid-ideally plastic half space by a flat rigid punch, up to the first order approximation as given by SPENCER /11/. An according though modified application to forging will follow below.

3.2 Elementary Kinetics of Non-Stationary Forging

We try to apply the perturbation method to plane-strain compression as shown in Fig.1, and use the notations of ch. 2.2. Preliminarily, K will again be assumed constant. However, instead of equ.(14) we will take the more realistic condition of COULOMB external friction,

$$|\tau| = -\mu\sigma_y \quad \text{at} \quad y = \pm H \tag{40}$$

into account. In case of small coefficients μ we even might assume that,

$$\mu = \varepsilon\mu' \; ; \quad \mu' \geqq 0 \tag{41}$$

which is a formal perturbation approach rather than a physical statement, and does never mean a linear rate dependence of friction! Because of equ.(41) the zero approximation to our problem just represents the static frictionless compression of the strip,

$$\left.\begin{aligned} &\underset{00}{\sigma_x} = 0 \,, \quad \underset{00}{\sigma_y} = -2K \,, \quad \underset{00}{\tau} = 0 \,, \\ &\underset{00}{u} = c\,\frac{x}{H} \,, \quad \underset{00}{v} = -c\,\frac{y}{H} \,, \end{aligned}\right\} \tag{42}$$

for which obviously the basic equations (13a-f) and the boundary conditions (15,19) hold. According to equ.(37) we have up to the first order approximation,

$$\left.\begin{aligned} &\sigma_x = \varepsilon \underset{10}{\sigma_x} + \beta \underset{01}{\sigma_x} \,, \qquad \sigma_y = -2K + \varepsilon \underset{10}{\sigma_y} + \beta \underset{01}{\sigma_y} \,, \\ &\tau = \varepsilon \underset{10}{\tau} + \beta \underset{01}{\tau} \,, \\ &u = c\,\frac{x}{H} + \varepsilon \underset{10}{u} + \beta \underset{01}{u} \,, \quad v = -c\,\frac{y}{H} + \varepsilon \underset{10}{v} + \beta \underset{01}{v} \,. \end{aligned}\right\} \tag{43}$$

After introducing the dimensionless quantities

$$\xi = \frac{x}{H} \,, \quad \eta = \frac{y}{H} \,, \quad s_x = \frac{\sigma_x}{2K} \,, \quad s_y = \frac{\sigma_y}{2K} \,, \quad \bar{\tau} = \frac{\tau}{2K} \tag{44}$$

we obtain by comparing both sides of equs.(13a,e,f) up to the first order terms in β and ε, regarding equs.(2),(4) the relations

$$\left.\begin{aligned} &|s_x - s_y| = 1 \,, \\ &\frac{\partial s_x}{\partial \xi} + \frac{\partial \bar{\tau}}{\partial \eta} = (2\varepsilon + \beta)\,\xi \,, \\ &\frac{\partial \bar{\tau}}{\partial \xi} + \frac{\partial s_y}{\partial \eta} = -\beta\eta \end{aligned}\right\} \tag{45}$$

the solution of which, unfortunately, fails like that given in ch.2.2 to fulfil the complete boundary conditions. Instead averageing them as before, we now pass to the well-established so-called elementary theory /2/ just by dropping vertical equilibrium (3 rd equ.(45)) but replace it by a linear η-distribution of shear stress so that the result following from equs.(40),(45),

$$\left.\begin{aligned} \bar{\tau} &= \mu\,\eta\, s_y, \quad \text{where} \quad s_y = s_x - 1\,, \\ s_x &= -\frac{2\varepsilon+\beta}{\mu}\left[\left(\frac{1}{\mu}-\xi\right) - \left(\frac{1}{\mu}-\delta\right) e^{\mu(\delta-\xi)}\right] \end{aligned}\right\} \qquad (46)$$

obeys strictly all the stress boundary conditions (15),(19) besides the realistic frictional law (40). This procedure is as experience shows, even better than having rigorous solutions under wrong boundary conditions.

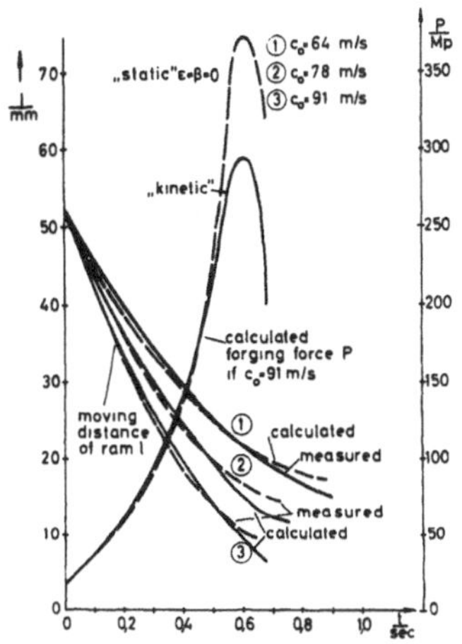

Fig.5. Upsetting of a free flying ram (impact velocity c_0) onto a cylindrical specimen (total height 2H = 52.5 mm, diameter D = 44.1 mm). Steel C 15, temp. ca. 1100 °C, friction μ = 0.15, mass of ram M = 1.04 kp s²/m⁴
Calculations: BEHRENS Experiments: ECKER .

Further items may be taken from literature /2,12/ where the results are additionally extended to sticking friction $|\tau|_{\eta=\pm1}=K$, as well as to strain hardening or even rate dependent materials. BEHRENS /13/ integrated over the time to get information on an entire forging process, and compared it with ECKER's experimental results /19/, Fig.5. The calculated "static" force for which formally ε = β = 0 has been adopted (i.e. ρ = 0) prooves to be a little smaller than the true kinetic one in the beginning since it need not store kinetic energy. This one, however, makes deformation easier at the end so that the top forces decrease by virtue of the inertia. The initial impact as such has not yet been considered but by means of the equations to be given in the final chapter it could be shown that the related maximum force at the very beginning, does not exceed the top given in Fig.5.

4. Further Non-Stationary Solutions

4.1 HAGEN-POISEUILLE Flow

Through the rigid circular tube, Fig.6, is streaming an incompressible plastic mass so that the physical shear stress $\tau = \sigma_{rz}$ and the associated strain rate $\chi = \frac{1}{2}\frac{dv}{dr}$ alone, are relevant while the normal stresses $\sigma = \sigma_r = \sigma_\psi = \sigma_z$ become equal and do not influence the process. $v = v(r,t)$ represents the only non-vanishing velocity. Kinetic equilibrium delivers us (cp./9/ and equ.(1))

$$\begin{aligned} &\frac{\partial\sigma}{\partial r} + \frac{\partial\tau}{\partial z} = 0 \,, \\ &\frac{\partial\sigma}{\partial z} + \frac{1}{r}\frac{\partial(r\tau)}{\partial r} = \rho\,\frac{\partial v}{\partial t} \,, \end{aligned} \tag{47}$$

and the yield condition must be

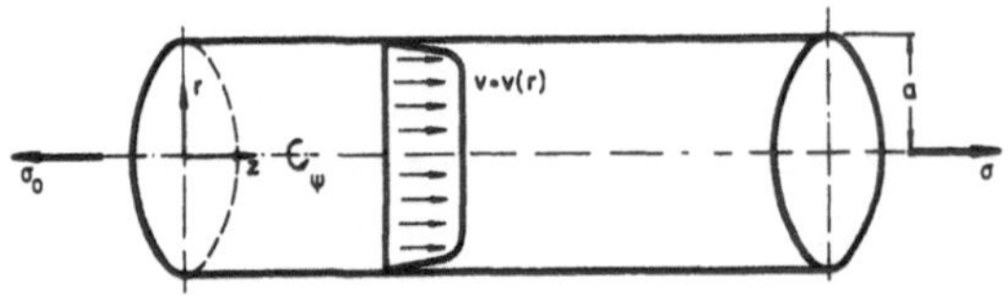

Fig.6. HAGEN-POISEUILLE flow

$$f = \tau + K = 0 \ . \tag{48}$$

It does not always allow for a continuous solution. So, if for instance K does not depend on the strain rate we obtain a rigid-body motion $v = v(t)$ with shear only occuring at the wall $r = a$ to which K refers in the following formulae,

$$\tau = -\frac{r}{a} K , \qquad \sigma = \sigma_o + \rho \dot{v} z + \frac{2}{a}\int_0^z K \, dz \ .$$

Then in the rigid kernel $0 \leqq r < a$, instead of equ.(48) only the condition of admissibility $f < 0$, cp. inequ.(8) holds. For a rate-dependent material, however, v may be determined from the pretty complicated differential equation which follows from equs.(47),(48),

$$\frac{1}{r}\,\frac{\partial^2 (rK)}{\partial v^2} - \frac{\partial^2 K}{\partial z^2} = \rho\,\frac{\partial^2 v}{\partial r \partial t} \ .$$

The above elementary consideration (cp./21/) has been presented as it is the only known, complete realistic three dimensional solution of plasto-kinetics which refers to rheology as well as to metal forming: Transport of a soft metal through a tube under high pressure.

4.2 Flexural Waves in Plate Bulging

In its initial state the circular blank (Fig.2, radius a, thickness $D_o = \text{const}$) must if not pre-stressed, never be considered as a membrane but as a plate for which we need examine small deformations only and therefore neglect geometric changes. $Q_1 = m_r$, $Q_2 = m_\psi$ denote bending moments per unit area ("couple stresses"), and $q^1 = \omega_r$, $q^2 = \omega_\psi$ the associated rates of curvature, Fig.7, so that equ.(5) holds, and

$$\omega_r = -\frac{\partial^2 v}{\partial r^2} , \qquad \omega_\psi = -\frac{1}{r}\,\frac{\partial v}{\partial r} , \tag{49}$$

(cp./14/) represent the compatibility conditions in which $v = v(r,t)$ is the velocity. Equilibrium may be taken from ref. /9/ (cp./14/) to give

$$\left.\begin{aligned} \frac{\partial\tau}{\partial r} + \frac{\tau}{r} + p &= 0 \ , \\ \frac{\partial m_r}{\partial r} + \frac{m_r - m_\psi}{r} - \tau &= 0 \ , \end{aligned}\right\} \qquad (50)$$

τ being the mean shear stress (Fig.7), $z = z(r,t)$ the deflection of the plate, and p comprising volume forces as well as

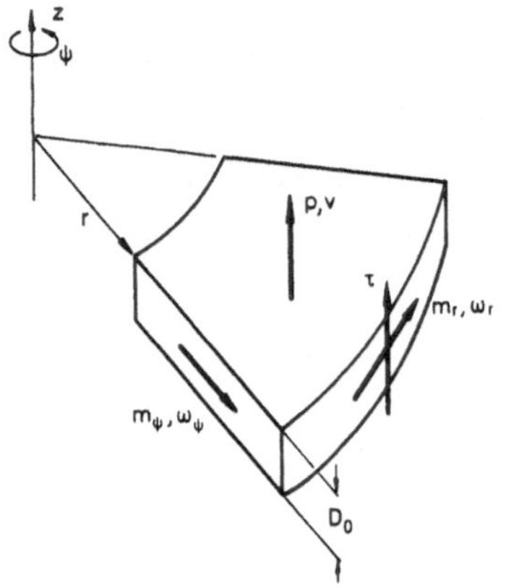

Fig.7. Notations for plate bending

external surface tractions in z-direction. We shall consider as in ch.2.3 a pure inertial motion so that p reduces according to Fig.2, to

$$p = -\rho\,\frac{\partial v}{\partial t} \ , \qquad v = \frac{\partial z}{\partial t} \qquad (51)$$

while membrane stresses (which would be initial pre-stresses only) are neglected.

The subsequent analysis starting from a classical paper of HOPKINS and PRAGER /14/ which was frequently extended as well as generalized (cp./3/), is less complicated than the approach given by WANG and HOPKINS /21/. At first we adapt TRESCA's yield condition (29) by integrating over the plate thickness, the stresses $\sigma_1, \sigma_2, \sigma_3$ which form a permutation of the stresses $\sigma_r, \sigma_\psi, \sigma_z \equiv 0$ radial, circumferential, and axial respectively to give the couple stresses

$$m_r = \frac{1}{D_o}\int_{-D_o/2}^{D_o/2} z\sigma_r dz \ , \quad m_\psi = \frac{1}{D_o}\int_{-D_o/2}^{D_o/2} z\sigma_\psi dz \ , \quad m_z = \frac{1}{D_o}\int_{-D_o/2}^{D_o/2} z\sigma_z dz \equiv 0$$

so that

$$f = m_1 - m_3 - N = 0 \quad \text{if} \quad m_1 \geqq m_2 \geqq m_3 , \tag{52}$$

where the material parameter N ("yield couple") results from integrating 2K sgnz . For simplicity we assume that N = const (ideally plastic material); $N = \frac{1}{2} K D_o^2$.

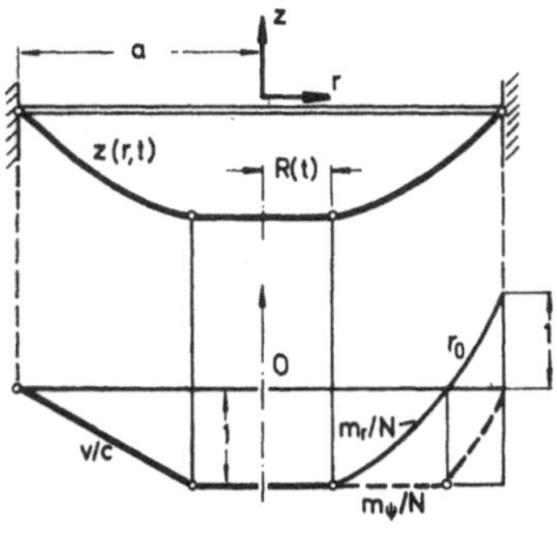

Fig.8. Inertial bulging of a circular plate built-in at its boundary, after initial impact. Circular yield hinges at r = R, r = a. Graph of the moments qualitative only

Now, one mode of deformation can be shown to be determined according to Fig.8, by a propagating circular "yield hinge" r = R(t)(flexural shock front in bending), and a stationary one at r = a in both of which alone, takes concentrated yielding place. In fact, the material velocity to be set up as follows,

$$v = \begin{cases} -c = \text{const} , & \text{if} \quad 0 \leqq r \leqq R(t) \\ -\dfrac{a - r}{a - R} c , & \text{if} \quad R(t) \leqq r \leqq a \end{cases} \tag{53}$$

is continuous at r = a (v ≡ 0) and r = R (v ≡ -c) but generates a local jump in curvature rate the sign of which demands because of (5) that $m_r \geqq 0$ at r = a, $m_r \leqq 0$ at r = R, while $\omega_\psi \leqq 0$ means $m_\psi \leqq 0$ for $R \leqq r \leqq a$. Thus regarding $m_z \equiv 0$, equs. (50),(51),(52), and (53) we obtain according to Fig.8 that

a) $m_r = m_\psi = -N$, $\tau = 0$, if $0 \leqq r \leqq R(t)$;

b) $\tau = -\frac{1}{6}\rho c a \frac{\dot{R}}{a} \frac{(R/a)^2}{(1-(R/a))^2} \frac{R}{r}\left\{-2\left(\frac{r}{R}\right)^3 + 3\frac{a}{R}\left(\frac{r}{R}\right)^2 + \left(2 - 3\frac{a}{R}\right)\right\}$,

if $R(t) \leqq r \leqq a$;

or specificly:

b1) $m_\psi = -N \leqq m_r \leqq 0$, if $R \leqq r \leqq r_o$;

b2) $m_r \geqq 0 \geqq m_\psi$, if $r_o \leqq r \leqq a$.

Using the boundary conditions $m_r = -N$ at $r = R$ for (b1), and $m_r = N$ at $r = a$ for (b2) the yield conditions $m_\psi = -N$ or $m_r - m_\psi = N$ respectively give us together with equilibrium (50) that

$$\frac{m_r}{N} = -1 - \frac{1}{12} \frac{\rho c a^2}{N} \frac{\dot{R}}{a} \frac{(R/a)^3}{(1-(R/a))^2} \frac{R}{r}\left\{-\left(\frac{r}{R}\right)^4 + 2\frac{a}{R}\left(\frac{r}{R}\right)^3 + \left(4 - 6\frac{a}{R}\right)\frac{r}{R} + 4\left(\frac{a}{R} - 1\right)\right\} \quad \text{if } R \leqq r \leqq r_o ,$$

$$\frac{m_r}{N} = 1 - \ln\frac{r}{a} - \frac{1}{6} \frac{\rho c a^2}{N} \frac{\dot{R}}{a} \frac{(R/a)^3}{(1-(R/a))^2} \left\{-\frac{2}{3}\frac{R}{a}\left(\frac{r}{R}\right)^3 + \frac{3}{2}\left(\frac{r}{R}\right)^2 + \left(2\frac{R}{a} - 3\right)\ln\frac{r}{a} - \frac{5}{6}\left(\frac{a}{R}\right)^2\right\} \quad \text{if } r_o \leqq r \leqq a .$$

The two missing quantities r_o/R, $\dot{R}/R$ must numerically be calculated from the condition that both above expressions disappear if $r = r_o$. They depend on R/a themselves so that $\dot{R}/R = g(R/a)$ represents a non-linear first order differential equation for $R = R(t)$ to be solved numerically itself. Items may be taken from the paper by WANG and HOPKINS /21/.

Let us merely add that the dynamically moving yield hinge is no more than a flexural wave - even transporting a discontinuity so that it is called "shock" wave or shock front (as it is an isolated one). Waves are characteristic for the first stage of deformation because any region of the body becomes plastic not before a first wave front had passed through it.

In general one may expect a continuous wave field while shock fronts in this example have been a consequence of simply assuming N constant. Thus they do not have a physical significance else which situation is different from, say, gas dynamics.

4.3 Uniaxial Longitudinal Waves in Upsetting

A rigid ram (mass M, velocity $v_o = v_o(t)$) shall impact at time $t_o = 0$ the infinite rod as shown in Fig.9 (initially constant cross sectional area A_o) in which only unaxial longitudinal waves (propagational velocity c) are examined. $v = v(x,t)$ leading especially to $v_o = v(0,t)$ denotes the longitudinal material velocity in convected, originally cartesian co-ordinates x to which the above wave velocity is related. v, however,

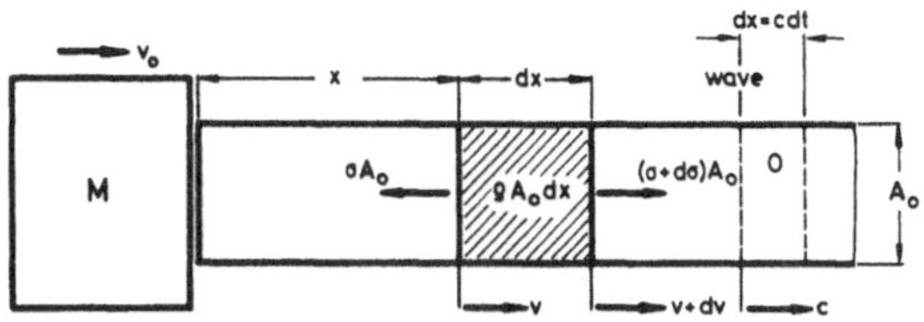

Fig.9. Longitudinal impact of a rigid ram onto an infinite rod

results from the longitudinal displacement $u = u(x,t)$ measured in the original cartesian coordinates x fixed in space, by

$$v(x,t) = \partial u(x,t)/\partial t \tag{54a}$$

while

$$\phi(x,t) = \partial u(x,t)/\partial x \tag{54b}$$

denotes the conventional, linear strain measure. Under conditions of loading, the constitutive equation

$$\sigma = \sigma(\phi) \tag{55}$$

of a rate-independent material refers here (in plasticity: exceptionally) to conventional stress which is related to the initial cross section A_o but may be calculated from the true physical stress σ_x (plotted in a yield curve), by means of

$$\sigma = \frac{\sigma_x}{1 + \varepsilon} \tag{56}$$

in case of an incompressible material ρ = const.

Thus Fig.10 shows graphs of equ.(55) as an example (/15,16/ transformed using equ.(56)[1]) from which we recognize that in the elastic range as well as in the plastic one, the conventional yield curves may be replaced by straight lines each which intersect at a fictitions yield limit

$$\phi = \tilde{\phi} \,, \qquad \sigma = \tilde{\sigma} \tag{57}$$

This holds approximately either for small or for large deformations showing different inclinations of the straight lines, respectively.

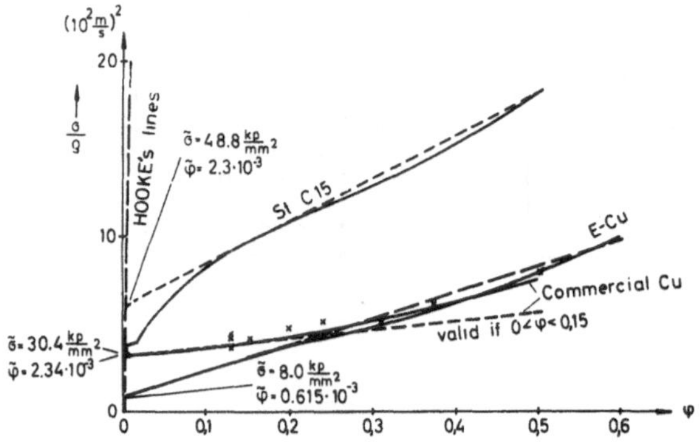

Fig.10. Conventional yield curves
——— measured (BÜHLER and SCHACK [steel], KRAUSE [E-copper])[1],
– – – linear substitution

Now from the dynamic equilibrium at the hatched element of Fig.9,

$$\frac{\partial \sigma}{\partial x} = \rho \frac{\partial v}{\partial t} \tag{58}$$

(ρ: initial density), and equs.(54a,b) we obtain because of $\partial\sigma/\partial x = (d\sigma/d\phi)(\partial\phi/\partial x)$ that

$$\left.\begin{aligned} \frac{\partial v}{\partial t} - c^2 \frac{\partial \phi}{\partial x} &= 0 \,, \\ \frac{\partial v}{\partial x} - \frac{\partial \phi}{\partial t} &= 0 \end{aligned}\right\} \tag{59}$$

---

[1] Author is indebted to Mr. J.ZACHMANN and his staff for performing the experiments using a commercial copper.

where

$$c = \left|\sqrt{\frac{d\sigma/d\phi}{\rho}}\right| \qquad (60)$$

makes sense if $d\sigma/d\phi > 0$ which is in general true.

For the elastic range, $c = c_e$, we obtain using YOUNG's modulus E that

$$c_e = |\sqrt{E/\rho}|, \qquad (61)$$

which is the well-known formula for noise propagation and delivers for steel ($\rho = 800$ kp s^2/m^4, $E = 21 \cdot 10^3$ kp/mm^2) or copper ($\rho = 910$ kp s^2/m^4, $E = 13 \cdot 10^3$ kp/mm^2) that $c_e = 5130$ m/s or $c_e = 3780$ m/s respectively. Even in the plastic range, using the linear approximations in Fig.10, becomes $c = c_p$ a constant, namely $c_p = 500$ m/s (steel), $c_p = 378$ m/s (E-copper), or $c_p = 220$ m/s (commercial copper, if $0 \leqq \phi \leqq 0.15$) respectively.[1]

If $c = $ const then from equs.(59) ϕ may be eliminated to give the wave equation

$$\frac{\partial^2 v}{\partial t^2} = c^2 \frac{\partial v}{\partial x^2} \qquad (62)$$

which confirms us that c means a wave velocity.

Let us observe now a wave front of length $dx = c\,dt$ as seen from the point O fixed in space (Fig.9), moving to the right. One measures the variation due to time, which is equal to the according negative ascent across the wave front, $\partial v/\partial t = -c\partial v/\partial x$ or $\partial\phi/\partial t = -c\partial\phi/\partial x$ respectively. Then regarding equs. (59),(60) and examining shock fronts only when differentials d sum up to finite differences Δ we obtain,

$$\Delta v = -c\,\Delta\phi \quad , \quad \Delta\sigma = \rho c^2\,\Delta\phi \qquad (63)$$

which two relations will govern our problem.

In front of the first elastic wave there is still the undisturbed state, $\phi_o = v_o = \sigma_o = 0$. Behind it in the elastic region, equs.(63) deliver us

$$\sigma_e = \rho c_e^2\,\phi_e = -\rho c_e v_e \,.$$

---

[1] c_e is lowered by three-dimensional effects, e.g. [23] . The same could be expected for c_p to hold.

As long as $|\sigma_e| \leqq \tilde{\sigma}$, $|\phi_e| \leqq \tilde{\phi}$ (Fig.10) or accordingly $v_e \leqq \tilde{v}$, where

$$\tilde{v} = \frac{\tilde{\sigma}}{\rho c_e} = c_e \tilde{\phi} \tag{64}$$

denotes some critical material velocity, there cannot occur any subsequent plastic deformation so that the velocity v_o of rams which are to forge must obey the inequality

$$v_o > \tilde{v} \,. \tag{65}$$

Considering the examples of Fig.10 we find e.g.

$\tilde{v}$ = 11.8 m/s for steel C15,
$\tilde{v}$ = 2.32 m/s for E-copper, and
$\tilde{v}$ = 8.85 m/s for commercial copper in case of small deformations.

Immediately behind the plastic shock front we obtain from equs.(63) regarding $v_e = \tilde{v}$, $\sigma_e = -\rho c_e \tilde{v}$, $\phi_e = -\tilde{v}/c_e$ that

$$\left.\begin{aligned} \phi_p &= \phi_e - \frac{1}{c_p}(v_p - v_e) = -\left[\frac{\tilde{v}}{c_e} + \frac{v_p - \tilde{v}}{c_p}\right], \\ \sigma_p &= \sigma_e - \rho c_p (v_p - v_e) = -\rho\left[c_e \tilde{v} + c_p (v_p - \tilde{v})\right] . \end{aligned}\right\} \tag{66}$$

As $|\phi_p|$ must not exeed 1 in which limit case the plastic zone of the rod shrinks to zero length, $v_o = v_p$, we find

$$v_o < \tilde{v}\left(1 - \frac{c_p}{c_e}\right) + c_p \,. \tag{67}$$

Now the velocity $v_o(t)$ of a free flying ram will decrease so that unloading occurs in the region behind the plastic front. This becomes unfortunately a mathematically rather complicated problem so that following an idea of LEE and WOLFE /18/,cp./13/, the unloading zone shall be considered rigid. In this case the elastic energy stored by the elastic shock front will never be paid out again but it is for forging negligeable as compared with the plastic work. Assuming therefore $v_o(t) = v_p$, as well as NEWTON's law of motion for the entire rigid region (mass $\rho A_o c_p t$) including the ram (mass M) which is loaded at the plastic front by $|\sigma_p| A_o = -\sigma_p A_o$ we obtain using equs.(66) the ordinary differential equation

$$(M + \rho A_o c_p t)\, \dot{v}_o(t) = \sigma_p A_o = -\rho A_o \left[c_e \tilde{v} + c_p \left(v_o(t) - \tilde{v}\right)\right] .$$

It possesses the integral

$$\frac{v_o(t)}{\tilde{v}} = \frac{1}{1 + \frac{c_p}{H} t} \left\{ \frac{v_o(0)}{\tilde{v}} - \left(\frac{c_e}{H} - \frac{c_p}{H}\right) t \right\}$$

where $v_o(0)$ is the velocity of impact, and

$$H = \frac{M}{\rho A_o}$$

the fictions length of the ram which it would have after being manufactured as a rod of cross section A_o equal to that of the specimen. The ram velocity as well as because of equ.(66), the plastic deformation $|\phi_p|$ in the shock front and the forging force $-M\dot{v}_o$ become maximal at the beginning but decrease afterwards so that the impact top of the forging force could be calculated. The process is over if $\phi_p = 0$, $v_o(t_p) = \tilde{v}$ which gives us the time $t_p = \frac{H}{c_e}\left(\frac{v_o(0)}{\tilde{v}} - 1\right)$, and the plastically deformed length of the rod l_p according to

$$\frac{l_p}{H} = \frac{c_p t_p}{H} = \frac{c_p}{c_e}\left(\frac{v_o(0)}{\tilde{v}} - 1\right) . \tag{68}$$

BEHRENS /13/ examined additionally the rod of finite length l but without reproducing his results (not even based on the simplified straight-line model in Fig.10 so that instead of shock fronts he also considered continuous wave fields), one may state that if $l_p/l < 1$ the deformation will take place only in a part of the specimen which thus becomes funnel-shaped. For larger values of l_p/l, also reflected waves from the end get influence. If one single reflection takes place so that the reflected wave deforms the other part of the specimen then we expect somewhat like a double-funnel or a vase while still larger values of l_p/l lead to multi-reflections and thus to a complete plastification so that the elementary theory of ch.3.2 (cylindrical or ton-like shape of the specimen)might be applied. Fig.11 shows according experimental results produced by means of a cartridge tool, using commercial copper (cp. Fig.10 under the assumption that $0 \leqq |\phi| \leqq 0.15$).

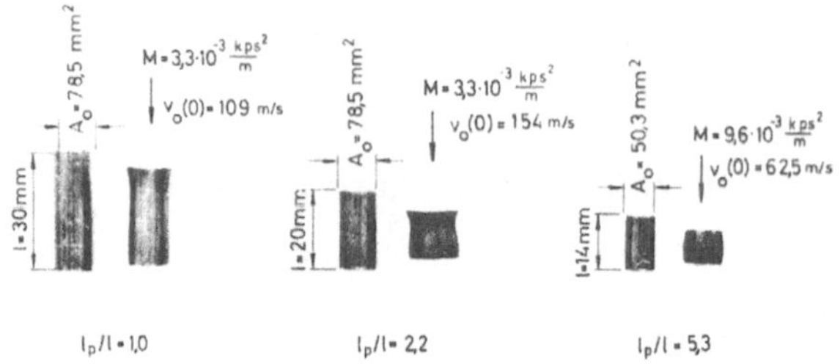

Fig.11. Deformation modes caused by plastic wave propagation in commercial copper

For the discovery of the mechanics of unaxial plastic waves, credit is used to be given to RAKHMATULIN (Russia), TAYLOR (England), and v.KARMAN (USA) who obtained their results during the war (1939-1945) but published with delay. MAZZOLENI's name, however, is hardly mentioned though, or because he examined at the same time the very civil process of forging only /20/. Further references may be taken from /3/, especially for rate dependent materials which strangely do not allow for plastic waves in the proper sense (cp./17/).

References

1. KOITER,W.T.: General Theorems for Elastic-Plastic Solids. Progr.Solid Mech. 1, 167-221 (1960)
2. LIPPMANN,H. und O.MAHRENHOLTZ: Plastomechanik der Umformung metallischer Werkstoffe. Vol.1, Berlin-Heidelberg-New-York: Springer 1967
3. CRISTESCU,N.: Dynamic Plasticity. Amsterdam: North Holland Publ.Comp. 1967
4. JOHNSON,W.: Impact Strength of Materials. London: Edward Arnold Publ. 1972
5. NAJAR,J.: Plane Polar-like Rapid Flow Problems for Perfectly Plastic Materials. J.Mécanique 7, 249-279 (1968)

6. BYKOVCEV,G.I.: Über das Pressen einer plastischen Schicht zwischen starren rauhen Platten unter Berücksichtigung der Trägheitskräfte. Jzv. Akad. Nauk SSSR Otd. techn. Nauk Mech. Masinostr. 6, 140-142 (1960) [Russian]
7. NAJAR,J.: Inertia Effects in the Problem of Compression of a Perfectly Plastic Layer Between Two Rigid Plates. Arch. Mech. Stosow. 19, 129-149 (1967)
8. LIPPMANN,H.: Kinetics of the Axisymmetric Rigid-Plastic Membrane Subject to Initial Impact. Int. J.Mech.Sci. 16 (1974), to appear
9. PARKUS,H.: Mechanik der festen Körper. Wien: Springer 1966 (2nd ed.)
10. JOHNSON,W., POYNTON,A., SING,H. and F.W.TRAVIS: Experiments in the Underwater Explosive Stretch Forming of Clamped Circular Blanks. Int. J. Mech. Sci. 8, 237-270 (1966)
11. SPENCER,A.J.M.: The Dynamic Plane Deformation of an Ideal Plastic-Rigid Solid. J. Mech. Phys. Solids 8, 262-279 (1960)
12. LIPPMANN,H.: Zur Dynamik des Schmiedens. Arch. Eisenhüttenwes. 35, 507-515 (1964); english version: On the Dynamics of Forging. Advances in Machine Tool Design and Research. Oxford-New York: Pergamon Press 1967, pp.53-66
13. BEHRENS,A.: Beitrag zur Theorie des Hochgeschwindigkeitsschmiedens. Diss. Braunschweig: Techn. Univ. 1968
14. HOPKINS,H.G. and W.PRAGER: On the Dynamics of Plastic Circular Plates. Z. angew. Math. Phys. 5, 317-330 (1954)
15. BÜHLER,H. und J.SCHACK: Formänderungsfestigkeit von Stählen im Bereich der Blauwärme. Draht-Welt 52, 489-495 (1966)
16. KRAUSE,U.: Vergleich verschiedener Verfahren zur Bestimmung der Formänderungsfestigkeit bei der Kaltumformung. (I) Arch. Eisenhüttenwes. 31, 745-754 (1963); (II) Stahl und Eisen 83, 1626-1640 (1963)
17. LIPPMANN,H. und A.BEHRENS: Zur Theorie elastisch-plastischer Wellen in dünnen Stäben. Z. angew. Math. Phys. 17, 62-68 (1966)

18. LEE,E.H. and H.WOLFE: Plastic Wave Propagation Effects in High Speed Testing. Trans. ASME J. Appl. Mech. 18, 379-386 (1951)

19. ECKER,W.: Ein Beitrag zum Gesenkschmieden unter Hämmern insbesondere hinsichtlich der Kraftmessung. Diss. Hannover: Techn. Univ. 1967

20. MAZZOLENI,F.: Das Schmieden von Metallen unter dem Hammer [in Italian]. La Metallurgia Italiana 35, 182-202 (1943)

21. WANG,A.J. and H.G.HOPKINS: On the Plastic Deformation of Built-in Circular Plates under Impulsive Load. J.Mech. Phys. Solids 3, 22-37 (1954)

22. FREDRICKSON,A.G.: Principles and Applications of Rheology. Englewood Cliffs, N.J.: Prentice-Hall 1964

23. LOVE,A.E.H.: Lehrbuch der Elastizität. Leipzig: B.G.Teubner 1907

AUTHOR INDEX

SUBJECT INDEX

LIST OF PUBLICATIONS

of

HEINZ PARKUS, Vienna

Monographs and textbooks

1. Wärmespannungen (with E.Melan) Wien: Springer. 1953
2. Instationäre Wärmespannungen. Wien: Springer. 1959
3. Mechanik der festen Körper. 2nd Ed., Wien: Springer.1966
4. Thermoelasticity. Waltham-Toronto-London: Blaisdell. 1968
5. Random Processes in Mechanical Sciences. Wien: Springer. 1969, (CISM)
6. Variational Principles in Thermo- and Magneto-Thermomechanics. Udine: CISM. 1970
7. Optimal Filtering. Wien: Springer. 1971, (CISM)
8. Magneto-Thermoelasticity. Wien: Springer. 1972, (CISM)

Articles in Handbooks

9. Civil Engineering Reference Book (Ed.J.Comrie), Chapters "Mechanics", "Elasticity" (with E.Chwalla) and "Vibrations". 2nd Ed., London: Butterworths. 1961
10. Handbook of Engineering Mechanics (Ed.W.Flügge), Chapter "Thermal Stresses". New York: McGraw-Hill. 1962

Papers

1946/47

11. Zur Stabilität des Einrotor-Hubschraubers. Oesterr.Ing.-Arch. 1, 58 (1946)
12. Drillschwingungen von Luftschraubenblättern. Oesterr.Ing.-Arch. 1, 296 (1947)

1948

13. The Disturbed Flapping Motion of Helicopter Rotor Blades. J.Aeronautical Sciences 15, 103 (1948)
14. Der wandartige Träger auf drei Stützen. Oesterr.Ing.-Arch. 2, 185 (1948)
15. Die Torsion geschlitzter Hohlwellen. Oesterr.Ing.-Arch. 2, 372 (1948)
16. Zur Berechnung der Luftkräfte an einer schwingenden Tragfläche. Anzeiger Oesterr. Akademie der Wissenschaften 10, (1948)

1949

17. Die spannungsoptische Methode und ihre Bedeutung für den Maschineningenieur. Maschinenbau u. Wärmewirtschaft 4, 77 (1949)
18. Beanspruchung und Schwingungen von Pleuelstangen. Oesterr. Ing.-Arch. 3, 222 (1949)
19. Die Torsion der Kreiswelle mit rechteckiger Längsnut. Oesterr.Ing.-Arch. 3, 336 (1949)

1950

20. Die überkritische Unterschallströmung. Oesterr.Ing.-Arch. 4, 88 (1950)
21. Ueber eine Erweiterung des Hamilton'schen Prinzipes auf thermoelastische Vorgänge. Federhofer-Girkmann-Festschrift, p.295. Wien: Deuticke. (1950)
22. Die Grundgleichungen der Schalentheorie in allgemeinen Koordinaten. Oesterr.Ing.-Arch. 4, 160 (1950)

23. Zur Berechnung von Pleuelstangen. Maschinenbau und Wärmewirtschaft 5, 152 (1950)
24. Calculation of Stresses in Connecting Rods. The Engineers Digest 11, 387 (1950)
25. Fundamentschwingungen von Schnellschlaghämmern. Oesterr. Bauzeitschrift 5, 123 (1950)

1951

26. Die Grundgleichungen der Schalentheorie in allgemeinen Koordinaten. Oesterr.Ing.-Arch. 6, 76 (1951)
27. Das Prinzip von Castigliano bei wärmebeanspruchten Körpern. Oesterr. Bauzeitschrift 6, 89 (1951)
28. Zur Berechnung von Gewölbestaumauern als Schalen. ZAMM 31, 277 (1951)
29. Die Grundgleichungen der allgemeinen Zylinderschale. Oesterr.Ing.-Arch. 6, 30 (1951)
30. Schweißspannungen in einer drehsymmetrischen Scheibe. Alfons Leon-Gedenkschrift, p.65. Wien: Verlag Allgemeine Bauzeitung. 1951
31. Wärmespannungen in Rotationsschalen mit drehsymmetrischer Temperaturverteilung. Sitzungsberichte Oesterr. Akademie der Wissenschaften 160, 1 (1951)

1953/54

32. Thermal Stress in Pipes. J.Appl.Mech. 20, 485 (1953)
33. Das Anlaufen einer Schubdüse mit vorgeschaltetem Rohr. Oesterr.Ing.-Arch. 8, 185 (1954)
34. Stress in a Centrally Heated Disk. Proceedings Second U.S. National Congress of Applied Mechanics, p. 307 (1954)

1955/56

35. Membranspannungen in der schiefen Kreiskegelschale. Oesterr.Ing.-Arch. 9, 196 (1955)
36. Periodisches Temperaturfeld im Keil. Oesterr.Ing.-Arch. 10, 241 (1956)
37. Zur Stabilität des Hubschraubers. Jahrbuch der Wissenschaftl. Gesellschaft f. Luftfahrt, p.55. Braunschweig: 1956

1959/60

38. Spannungen beim Abkühlen einer Kugel. Ing.-Arch. 28, 251 (1959)

39. Elastic Thermal Stresses in Delta Wings. Part I. AFOSR, TR 60-140 (1960)

1961/62

40. Die Schlagbewegung eines Hubschrauber-Rotors in turbulenter Luft. ZFW 9, 217 (1961)

41. Temperaturfelder bei zufallsabhängiger Oberflächentemperatur. Anzeiger Oesterr. Akademie der Wissenschaften, 153 (1961)

42. Wärmespannungen bei zufallsabhängiger Oberflächentemperatur. ZAMM 42, 499 (1962)

43. Elastic Thermal Stresses in Delta Wings. Part II. AFOSR, TR 60-140 (1962)

1963

44. Die "Durchschlagszeit" einer viskoelastischen flachen Kugelschale. Oesterr.Ing.-Arch. 17, 165 (1963)

45. Methods of Solution of Thermoelastic Boundary Value Problems. Proceedings Third Symposium on Naval Structural Mechanics, p.317, New York: 1963, (Invited lecture)

1964/65

46. Wärmespannungen. III.Lehrgang für Raumfahrttechnik. Bd.III/309 ff. Aachen: 1964

47. Der Einfluß von Eigenspannungen auf die Torsion dünnwandiger offener Profile (with E.Tungl). Stahlbau und Baustatik, aktuelle Probleme. p.135 ff. Wien: Springer. 1965

1966

48. Stability of Elastic Plates under Random Membrane Stress. Bull.Acad.Pol.Sc., Serie sc.techn. 14, 93 (1966)

49. Grundlagen und Probleme der Thermo- und Viskothermoelastizität. ZAMM 46, T 16 (1966), (Hauptvortrag)

1967

50. On the Lifetime of Viscoelastic Structures in a Random Temperature Field. Recent Progress in Applied Mechanics. p.391, Stockholm: Almqvist and Wiksell. New York: Wiley. 1967
51. Einige stochastische Eigenschaften des linearen Schwingers (with J.L.Zeman). ZAMM 47, T 96 (1967)
52. Stochastische Störungen optimaler Raketenbahnen (with F.Ziegler). ZAMM 47, T 97 (1967)
53. A Stochastic Problem of Thermoelasticity (with J.L.Zeman). Proc.VIIIth Euromech-Symposium, p.171, Warsaw: 1967

1969

54. Stochastic Perturbations of Optimal Rocket Trajectories (with F.Ziegler). Proceedings 2nd IFAC-Symposium on Automatic Control in Space, p.699, Pittsburgh: Instrument Soc. of America. 1969
55. Thermal Effects in Viscoelastic Structures. Proceedings 2nd Canadian Congress of Applied Mechanics, p.455, Waterloo: 1969. (Invited lecture)

1970

56. Some Stochastic Problems of Thermoviscoelasticity (with J.L.Zeman). Proc. IUTAM-Symposium on Thermoinelasticity. East Kilbride 1968, p.226, Wien-New York: Springer. 1970
57. Kriechknicken bei langsam pulsierender Axialkraft. Klaus-Oswatitsch-Festschrift. Wien: Techn. Hochschule, Inst. f. Strömungslehre. 1970
58. Note on the Behavior of Thermorheologically Simple Materials in Random Temperature Fields (with H.Bargmann). Acta Mechanica 9, 152 (1970)

1971/72

59. Constitutive Equations of the Linear Viscoelastic Dielectric. Trends in Elasticity and Thermoelasticity,

p.191, Groningen: Wolters-Noordhoff. 1971

60. Beulen einer Platte im transversalen Magnetfeld bei fluktuierender Temperatur. ZAMM 52, T 143 (1972)

61. Die Beulwahrscheinlichkeit einer Platte in einem transversalen Magnetfeld. Sitzungsberichte Oesterr.Akademie d. Wissenschaften, Math.-naturw.Klasse, 180, 185 (1972)

62. Thermoelastic equations for ferromagnetic bodies. Archives of Mechanics 24, 819 (1972)

63. Magneto- und Elektroelastizität. ZAMM 53, T 18 (1972)

1974

64. Der Glättungseffekt bei erdverlegten Rohren unter Innendruck (with H.Troger). Oesterr.Ing.Z. 17, 87 (1974)

65. Grundgleichungen elastischer anisotroper Medien und Lösungsbeispiele. Anisotropie-Symposium Wien 1973. Wien-New York: Springer, (im Druck)

Proceedings and Journals

Seminar in Thermal Mechanics. Michigan State University. East Lansing: 1959

Irreversible Aspects of Continuum Mechanics (with L.I.Sedov). IUTAM-Symposium, Vienna 1967. Wien-New York: Springer. 1968

Oesterr.Ing.-Arch. 1959-1964. Wien: Springer

Acta Mechanica (with A.Phillips). Wien-New York: Springer. Since 1965